PTP

The ASNT PERSONNEL TRAINING PUBLICATIONS

PT

Liquid Penetrant Testing Classroom Training Book

Written for ASNT by

Duane Badger
NDT Assistance

The American Society for Nondestructive Testing, Inc.

Published by The American Society for Nondestructive Testing, Inc.
1711 Arlingate Lane
Columbus, OH 43228-0518

ASNT exists to create a safer world by promoting the profession and technologies of nondestructive testing

ISBN 1-57117-121-5

Printed in the United States of America

Library of Congress Cataloging-in-Publication Data

Liquid penetrant testing : classroom training book / written for ASNT by Duane Badger.
p. cm. -- (Personnel training publications series)
Includes bibliographical references and index.
ISBN 1-57117-121-5 (pbk.)
1. Penetrant inspection. I. American Society for Nondestructive Testing. II. Title. III. Series.

TA417.55.B33 2004
620.1'127--dc22

2004028159

First printing January 2005

Acknowledgments

A special thank you goes to the following technical reviewers who helped with this publication:

James Borucki, Gould Bass NDT
Gina Caudill, United States Coast Guard
Nat Faransso, KBR
Art Gallant, Elite Inspection Services
Vilma Holmgren, Magnaflux
Terry Kessler, TK Consulting, LLC
Brian MacCracken, Pratt and Whitney
Joe Mackin, International Pipe Inspectors Association
David Moore, Sandia National Laboratories
William Mooz, Met-L-Chek Company
Jerry Nelson
Sam Robinson, Sherwin, Inc.
James Waugh, Global Technical Services

special thank you goes to the following corporate reviewers who helped with this publication:

John Brausch, Air Force Research Laboratory
Jerry Fulin, Tennessee Gas Pipeline
Jim Houf, The American Society for Nondestructive Testing, Inc.
Scott Huddleston, Huddleston Technical Services
Stephen Lakata, Norfolk Southern Corporation
S. O. McMillan, BAV Quality Assurance
Joseph E. Monroe, Eastern NDT
David R. Quattlebaum, Quattlebaum Consultants
Frank Sattler, Sattler Consultants, Inc.
Stephen Senne, Senne Technical Services
Noel Tracy, Universal Technology Corporation
Ron VanArsdale, Inspection Training and Consulting
Albert M. Wenzig, Jr., Industrial Testing Laboratory Services, LLC

e Publications Review Committee includes:

Chair, Sharon I. Vukelich, University of Dayton Research Institute
B. Boro Djordjevic, Johns Hopkins University
Mark A. Randig, Cooperheat-MQS, Inc.

n E. Spence
cational Materials Editor

Foreword

The American Society for Nondestructive Testing, Inc. (ASNT) has prepared this series of *Personnel Training Publications* to present the major areas in each nondestructive testing method. Each classroom training book in the series is organized to follow the Recommended Training Course Outlines found in *Recommended Practice No. SNT-TC-1A*. The Level I and Level II candidates should use this classroom training book as a preparation tool for nondestructive testing certification. An ASNT NDT Level I or Level II may be expected to know additional information based on industry or employer requirements.

Table of Contents

PTP
The ASNT
PERSONNEL TRAINING
PUBLICATIONS

PT

LEVEL I

Chapter 1

Introduction to Liquid Penetrant Testing

Liquid penetrant testing is a versatile nondestructive test method used for detecting surface discontinuities in a wide variety of solid, nonporous materials. The effectiveness of the test is determined by the training, skill and dedication of the penetrant technician, the cleaning and preparation of the test object and the materials and procedure used to perform the test. Liquid penetrant testing is used primarily for detecting surface discontinuities.

History

The oil and whiting method used in the railroad industry in the early 1900s was the first recognized use of the principles of penetrants to detect cracks. The oil and whiting method used an oil solvent for cleaning followed by the application of a whiting or chalk coating, which absorbed oil from the cracks revealing their locations. Soon a dye was added to the liquid. By the 1940s, fluorescent or visible dye was added to the oil used to penetrate test objects.

Experience showed that temperature and soak time were important. This started the practice of written instructions to provide standard, uniform results. The use of written procedures has evolved, giving the ability for design engineers and manufacturers to get the same high standard results from any properly trained and certified liquid penetrant testing technician.

The Purpose of Liquid Penetrant Testing

Liquid penetrant testing can detect surface discontinuities. It is a relatively inexpensive test method considering personnel training time, materials used and the ease with which surface discontinuities can be located. When performed properly with proper process controls and qualified technicians, it will provide reliable indications. Liquid penetrant testing is used to locate surface discontinuities on the final finished surface of test objects, but can also be useful for inservice checks to detect fatigue damage or production problems, for example the root pass of some welds and fatigue cracks. Fatigue induced cracks originate or propagate to the surface, making liquid penetrant tests useful in their detection.

Many materials fail during use. The cause of a high percentage of these failures is a linear discontinuity left in the test object during the manufacturing process or developed during use of the test object. This has convinced design engineers, manufacturers, users, safety and regulatory agencies to use reliable, inexpensive surface testing to minimize the undesirable consequences and expenses of test object failure. The goal of any inspection process is to achieve as high a probability of detection as possible to mitigate risk of part failure.

Basic Principles

When a dye is added to a liquid with a certain combination of properties of cohesion, adhesion, surface tension and viscosity, the liquid is called *penetrant*, *penetrant dye* or in some specifications just *dye*. When this penetrant is placed on a clean, dry surface it will wet the surface properly and migrate by capillary action. This capillary action causes the liquid to enter discontinuities that are open to the surface and flow up or down into the opening.

Some specifications and company procedures require different penetrant dwell times for different types of materials or discontinuities. These times will be specifically required for the particular test object or procedure referenced. Typical penetrant dwell times are between 5 and 45 min. This penetrant dwell time allows the capillary action to cause the liquid to enter discontinuities, as shown in Figure 1.1. After excess surface penetrant is removed, only the penetrant in the discontinuities

Figure 1.1: Penetration of liquid penetrant onto a surface discontinuity.

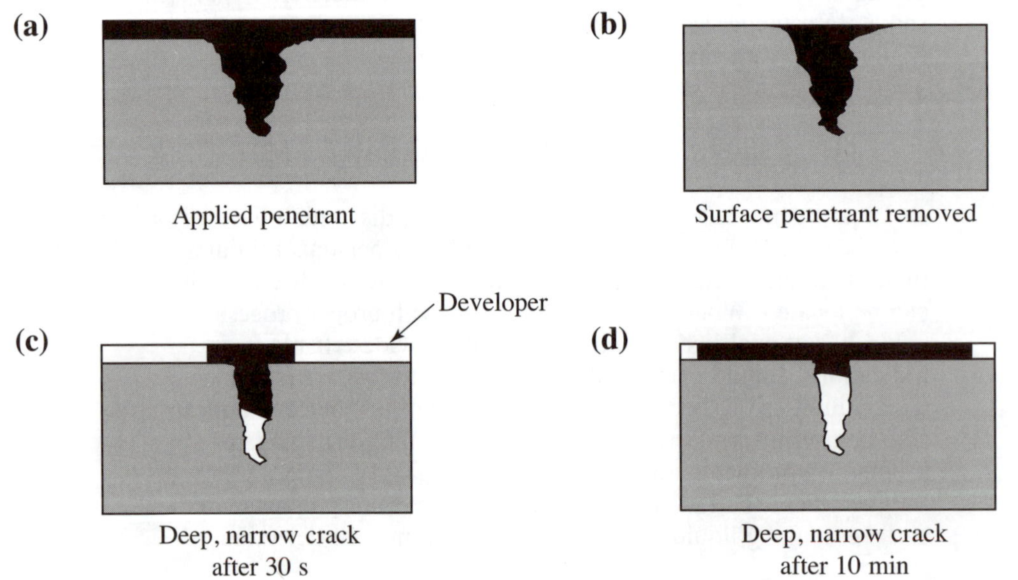

remains. Reverse capillary action causes the penetrant to migrate or bleed back out to form a penetrant indication.

The use of a contrasting developer aids the reverse capillary action bleedout and makes the indications easier to see. A liquid penetrant testing indication is the penetrant bleedout spot against the contrasting developer background. After a developer dwell time, which is usually also a minimum of 10 min, any liquid penetrant indications are evaluated to the specified acceptance criteria.

Developer dwell times vary with specification, company procedures or techniques; type of materials; type of discontinuity to be detected; and type of developer used. These dwell times typically range between 5 min and 4 h. Figure 1.2 illustrates the typical sequence of a liquid penetrant test.

Figure 1.2: Sequence of liquid penetrant tests. (Local or spot application shown here.) For process line application, immersion tanks or bulk spray applications are typically used.

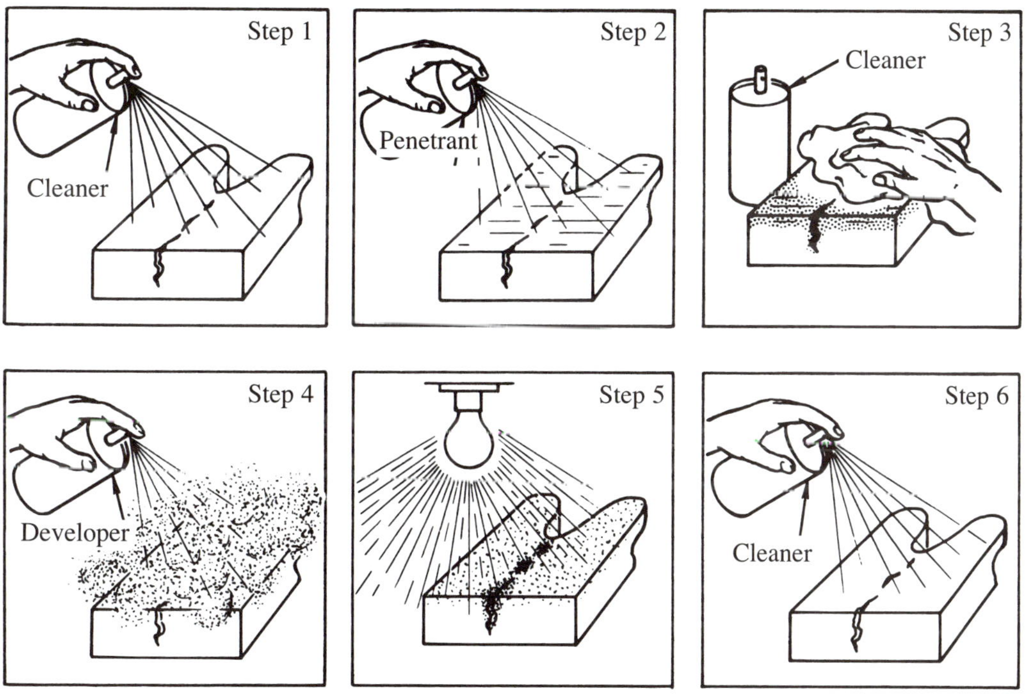

Step 1: The test object or spot is thoroughly cleaned and dried.
Step 2: The penetrant is applied. Dwell time is allowed to penetrate discontinuities.
Step 3: The excess surface penetrant is removed.
Step 4: Developer is applied
Step 5: Indications are evaluated and accepted or rejected.
Step 6: After the penetrant test, the test object or spot is postcleaned.

Commercially Available Liquid Penetrant Materials

The various combinations of liquid penetrant materials that are commercially available and commonly used are changing rapidly as new and better materials are formulated. The basic classification system remains fairly uniform and classifies the penetrant dye by type: fluorescent dye which is Type I, or visible dye which is Type II. The fluorescent penetrants are further classified into sensitivity levels: ultra low to ultra high. The different processes are then classified by the method used to remove the excess penetrant.

The developers are classified as Forms a through f. They include dry powder, water soluble, water suspendible, nonaqueous and specific application developers. The basic liquid penetrant classification system is shown in Table 1.1.

Table 1.1: Types and methods of liquid penetrant testing.

Type	Method	Sensitivity	Developer	Solvent
Type I: fluorescent penetrant	**Method A**: water washable	**Level 1/2***: ultra low	**Form a**: dry powder	**Class 1**: halogenated
Type II: visible penetrant	**Method B**: lipophilic postemulsifiable	**Level 1**: low	**Form b**: water soluble	**Class 2**: nonhalogenated
	Method C: solvent removable	**Level 2**: medium	**Form c**: water suspendible	**Class 3**: special application
	Method D: hydrophilic postemulsifiable	**Level 3**: high	**Form d**: nonaqueous Type I (for fluorescent)	
		Level 4: ultra high	**Form e**: nonaqueous Type II (for visible)	
			Form f: special application	

* Sensitivity level 1/2 applies to Type I, Method A penetrants only. There is no sensitivity level classification for Type II penetrant systems.

Personnel Qualification and Certification

The most widely used organization for nondestructive testing training and qualification guidelines is The American Society for Nondestructive Testing (ASNT). ASNT's recommended guideline is called *Recommended Practice No. SNT-TC-1A*. ASNT also issues the *ANSI/ASNT CP-189: Standard for Qualification and Certification of Nondestructive Testing Personnel*. ASNT certifies Level II and Level III personnel through the ASNT Central Certification Program (ACCP). ASNT certifies Level III personnel who are responsible for the qualification and certification of companies' nondestructive testing Level I and Level II technicians.

When a person starts the training and qualification process, they are classified as a trainee. Their goal is usually to become a fully certified Level II or Level III technician. Qualification is the organized training and required testing to prove the person has proper training. Certification is written testimony by the employer that the person is completely qualified to the company's written practice (as described in *Recommended Practice No. SNT-TC-1A* or *National Aerospace Standard NAS 410, NAS Certification and Qualification of Nondestructive Testing Personnel*).

To become qualified, a person must have the prescribed organized training and work experience under the supervision of a certified Level II or Level III technician. This work experience includes processing test objects as directed by the Level II technician, but never doing the evaluation for acceptance or rejection. The Level II technician supervises the processing so that if the trainee makes an error, the test can be repeated and the Level II technician can evaluate the test object and accept or reject, and the trainee will observe the evaluation.

After the company's required work experience, the trainee shall take written general and specific qualification tests and a practical test on at least two test objects, or one test object for each technique to be used. If the individual passes all tests, they can be certified as a Level I or Level II by the company. It is possible for all of the required organized classroom training and work experience to be accomplished as a trainee, and the person can be tested and certified directly to Level II.

All technicians are required to have a vision acuity test on an annual basis. This test can be taken either with or without corrective lenses. However, if corrective lenses are worn, they must be the same corrective lenses worn during liquid penetrant testing. A color contrast differentiation examination must also be given to demonstrate that the individual can distinguish between the colors used for liquid penetrant testing.

Employee Certification

Training

Training involves an organized program developed to provide test personnel with the knowledge and skill necessary for qualification in a specific area. This is performed in a classroom setting where the principles and techniques of the particular test method are reviewed. The length of training required is stated in the employer's written practice.

Experience

Experience includes work activities accomplished in a particular test method under the supervision of a qualified and/or certified individual in that particular method. This is to include time spent setting up tests, performing calibrations, specific tests and other related activities. Time spent in organized training programs does not count as experience. The length of experience required is stated in the employer's written practice.

Examination

Level I and Level II personnel shall be given written general and specific examinations, a practical examination and a visual examination. The general examination shall cover the basic principles of the applicable method. The specific examination shall cover the procedures, equipment and techniques that the employee will be required to perform in their job assignment. The practical (hands on) examination allows employees to demonstrate their ability to operate the appropriate test equipment and to perform inspections using that equipment in accordance with appropriate specifications. Level III personnel must pass written basic, method and specific examinations. Testing requirements are stated in the employer's written practice.

Certification

Certification of nondestructive testing personnel is the responsibility of the employer. Personnel may be certified when they have completed the initial training, experience and examination requirements described in the employer's written practice. The length of certification is stated in the employer's written practice. All applicants shall have documentation that states their qualifications according to the requirements of the written practice before certifications are issued.

Safety Precautions

Liquid penetrant testing uses a variety of materials that have some hazardous characteristics. Except for water, the removers and nonaqueous carriers are usually combustible, and with unprotected contact can cause skin irritation. The developing powders are nontoxic, but can be a health hazard in confined spaces. The ultraviolet lamps used with fluorescent penetrants can cause physiological damage. All of these hazards can be avoided or minimized by observing the precautionary measures in the material safety data sheets supplied with penetrant materials.

Fire

Flashpoint is the lowest temperature at which vapors above a volatile combustible substance ignite in air when exposed to flame. The higher the flashpoint of a material, the less fire hazard it presents.

Safe practice requires that penetrant materials used in open tanks have a minimum flashpoint of 93 °C (200 °F). Smoking is forbidden in or near test areas. Penetrant materials are never stored near heat or open flame, and exhaust fans are used to disperse vapors. Aerosol cans should never be sprayed near open flames, sparks or welding arcs.

Skin Irritation

The oil bases of liquid penetrant materials have a drying action on the skin. Because of this, the materials may cause unpleasant, if not dangerous, irritations. To prevent unnecessary contact with penetrant materials, care is taken to avoid splashing; protective hand creams are used; aprons, neoprene gloves and face shields are worn; and soap and water are used to immediately remove any penetrant materials that have come in contact with the skin.

Air Pollution

Dust and vapors from materials used in liquid penetrant testing are nontoxic, but inhalation of excessive amounts can be a health hazard. To avoid unhealthy concentrations of developer powder in the atmosphere, exhaust fans are installed in confined areas where dry developers are used. Fans are also used to remove the vapors from the test area. Always follow recommendations for respirator or mask use.

Ultraviolet Radiation

The ultraviolet radiation used to cause fluorescence of penetrant materials has a frequency of about 365 nm. This frequency is at the low end of the ultraviolet spectrum. Higher frequency ultraviolet rays are harmful to many forms of life including humans. The ultraviolet lamp filters used with liquid penetrant tests filter out most of the harmful ultraviolet rays generated by the mercury arc lamp, as well as the visible light rays.

Personnel should never look into the ultraviolet radiation. Wearing long sleeves and gloves when handling small test objects while the ultraviolet radiation source is on will help prevent any possible skin exposure. Missing, cracked or broken filters should be replaced before use. Protective goggles or lenses should be used as required or recommended.

Chapter 2

Liquid Penetrant Processing

Procedures and Techniques

All liquid penetrant testing required by a contract or specification is performed in accordance with a referenced or supplied standard procedure or technique. Procedures can be broad and cover several specific techniques. Techniques are usually a one page data sheet that provides enough information so that any qualified Level II liquid penetrant technician can perform or repeat a liquid penetrant test on a specific test object using a standard procedure.

Each procedure and technique shall be approved and signed by a certified Level III technician. In some cases, the approval certificate is submitted to the prime contractor or customer for approval, and then can be used for standard tests by a Level I or Level II certified technician.

Preparation of Test Objects

The properties of the test object determine the type of surface preparation that will be used. Nickel alloys, certain stainless steels and titanium have an affinity for specific elements (for example, sulfur or chlorine), and if exposed to them may become structurally damaged. In all cases, the test object must be free of any foreign material that will interfere with the test and must be completely dry.

Certain cleaning and finishing operations are prohibited before liquid penetrant tests, as they may smear metal and close the surface openings of discontinuities. These operations may include power wire brushing, grit blasting and shot peening, but can also include other metal smearing operations.

Typical industry specifications prohibit operations that could close or mask surface openings before liquid penetrant testing. Some specifications also require a specific method to remove any such smeared metal, such as an approved acid etching process. A smooth surface is preferred, but liquid penetrant testing can be satisfactorily performed on welded surfaces, relatively rough castings and a wide variety of surface finishes using the properly designed penetrant materials and proper techniques for excess penetrant removal.

Precleaning of Test Objects

There are many satisfactory precleaning methods depending on the test material and the condition of the surface of the test object. All cleaning methods must meet local environmental, health and safety requirements and cause no harm to the test object. Many liquid penetrant testing procedures have specific requirements for types of precleaning operations. As a minimum, all test objects are wiped clean with an approved solvent, and the solvent is allowed to dry completely.

Precleaning and Postcleaning Equipment

Proper cleaning is essential to liquid penetrant testing for two reasons.

1. If the test object is not physically and chemically clean and dry, penetrant testing may be ineffective.
2. If all traces of penetrant materials are not removed after the test, they may have a harmful effect after the test object is placed in service.

The cleaning processes commonly used with penetrant testing are discussed in the following sections. Unless the cleaning methods and chemicals are known to be compatible with the penetrant, a final cleaning should be done with the solvent recommended by the manufacturer.

Solvent Cleaning

Solvent cleaning may use tanks for immersion, or the solvent material may be spayed, brushed or wiped on and wiped off. Solvent cleaning is the process most commonly used for spot inspections. However environmental, health and safety concerns are making detergent cleaning and steam cleaning more attractive options. A solvent cleaner must evaporate readily and completely from the surface and from surface connected discontinuities. Solvent cleaners should only be used to remove organic contaminants.

Detergent Cleaning

Immersion tanks or and detergent solutions are common means of accomplishing the cleaning required by liquid penetrant tests. The detergents wet, penetrate, emulsify and saponify (change to soap) various soils. The only special equipment requirement imposed by penetrant test cleaning is the need for suitable rinsing and drying facilities. When thoroughly rinsed and dried, detergent cleaning leaves a test surface that is physically and chemically clean. Detergent cleaners should have a combination of detergency (cleaning) dispersion, emulsifying, foaming, solubilizing and wetting properties.

Vapor Degreasing

Vapor degreasing is effective in the removal of oil, grease and similar organic contamination. However, there are restrictions as to its use before and after liquid penetrant testing. Safety and environmental concerns have virtually eliminated vapor degreasing. Degreasing must be limited to those materials that have been approved for this method of cleaning. Vapor degreasing removes organic soils from both the surface and cracks, evaporates completely and, unlike water based cleaners, does not require a rinse step or a drying (oven) step.

Steam Cleaning

Steam cleaning equipment is particularly adaptable to the cleaning of large, unwieldy test objects not easily cleaned by immersion. No special equipment is required for steam cleaning of test objects destined for liquid penetrant testing. Steam with alkaline detergent provides ideal precleaning conditions. The alkaline detergent emulsifies, softens or dissolves the organic contaminant, and the steam gives the mechanical action to remove the alkaline detergent/contaminant from the test object.

Ultrasonic Cleaning

Ultrasonic agitation is often combined with solvent, detergent or alkaline cleaning to improve efficiency and reduce cleaning time. Ultrasonic cleaning equipment is useful in the cleaning of large quantities of small test objects. In many cases, special approvals must be granted to use ultrasonic agitation.

Rust and Surface Scale Removal

Any approved commercially available acid or alkaline rust remover may be used for precleaning. Required equipment and procedures are specified in the manufacturer's directions.

Paint Removal

Dissolving type hot tank paint strippers and bond release or solvent paint strippers may be used to remove paint in precleaning. Required equipment and procedures are as specified in the manufacturer's directions.

Etching

Test objects that had metal smearing operations, such as power wire brushing or sandblasting, often require etching to prepare them for liquid penetrant testing. This process uses an acid or alkaline solution to open up grinding burrs and remove smeared metal from surface discontinuities. All acid or alkaline residues must be neutralized and removed before liquid penetrant testing. The etching and neutralizing processes use either tanks and immersion or manual equipment and materials.

Precleaning Processes to be Avoided

Blast (shot, sand, grit or pressure), liquid honing, emery cloth, power wire brushes and metal scrapers should not be used on the test object before liquid penetrant testing. These processes tend to close discontinuities by smearing metal, peening or cold working the surface. A hand wire brush may be helpful in removing rust, surface scale or paint. Relatively fine bristle brushes should be used and light pressure exerted to prevent smearing of softer metals.

Precleaning and Postcleaning Materials

Specific test objects or systems may require special cleaning materials.

Drying Test Objects

A very important step in the liquid penetrant testing process is to ensure all tested surfaces and all potential discontinuities are completely dry. If any liquid precleaner remains in discontinuities, the penetrant may be unable to enter discontinuities and an inadequate exam will be performed. This is critical because the technician may not be aware that penetrant did not enter the discontinuities.

Some liquid penetrant testing procedures require test objects that are cleaned with water receive a final cleaning with a water soluble solvent, such as isopropyl alcohol or acetone, to promote evaporation of water from discontinuities. Some liquid penetrant testing procedures require that test objects are oven dried at temperatures up to 71 °C (160 °F). As a minimum, all surfaces should be allowed to dry by normal evaporation for a minimum of 5 min.

Dwell Time

The penetrant dwell time is the length of time that the penetrant is allowed to wet the surface and soak into discontinuities. The developer dwell time is the length of time that the penetrant is allowed to soak into the developer before the indications are evaluated.

Application of Penetrant

The application of penetrant is performed after the test surface is completely clean and has been properly dried. Almost any method is allowed for penetrant application, including spraying, brushing, pouring or dipping. Penetrant should never be allowed to dry completely on the test surface. The test object must be turned or moved to prevent pooling of penetrant during the dwell time. It is important that all test surfaces are completely wetted with a thin coat of penetrant for the entire specified dwell time. Use of an ultraviolet radiation source will assure that all surfaces remain wetted with penetrant and no water break areas occur where the penetrant film separates.

Removal of Excess Surface Penetrant

Adequate Illumination

Adequate illumination at the rinse station for fluorescent penetrant is about 100 lx (10 ftc) of white light and greater than 100 µW/cm^2 of near ultraviolet radiation (UV-A). This lighting allows better monitoring of removal of excess fluorescent penetrant and limits possible over rinsing or under rinsing.

The ultraviolet radiation background lighting at the rinse station also allows the technician to ensure complete coverage of the test object when using emulsifier, since the emulsifier is a different color than the penetrant under the ultraviolet radiation. The rinse station may be illuminated to the same level as the inspection booth (1000 μ/cm^2 and less than 20 lux [2 ftc]) if desired.

The evaluation of liquid penetrant indications requires proper illumination of the complete testing area. This is usually specified in the liquid penetrant testing procedure. For visible dye penetrant, 1000 lx (100 ftc) at the test surface is adequate. For fluorescent penetrant tests, the surface should be illuminated with 1000 µW/cm^2 of near ultraviolet radiation (UV-A) as measured at the test surface, and the white light should be less than 20 lx (2 ftc). These lighting levels will meet the requirements of most industrial specifications. However, verification of the proper illumination requirement should be given in the standard procedure or technique being used. Personnel in the testing area should not wear white clothing, which reflects large amounts of visible light when exposed to ultraviolet radiation.

Water Rinse (Method A, B and D)

After the required penetrant dwell time and proper emulsification (if applicable), the water rinse should be coarse droplets normally applied at an oblique angle from a distance of about 30 cm (12 in.). The standard nozzles available from approved nondestructive testing suppliers have been qualified for standard industry techniques. The preset pressure should not exceed 40 psi (275 kPa), and the temperature should be between 10 to 38 °C (50 to 100 °F). When hydro-air nozzles are used, the air pressure should not exceed 25 psi (172 kPa).

The rinse is typically accomplished at a rinse station with adequate illumination (ultraviolet radiation for Type I penetrant and white light for Type II penetrant). The light should shine on the test object so that the technician can evaluate when the excess penetrant is removed. The technician should also ensure there is no over washing. When the excess surface penetrant is removed, the surface water is drained or removed from cavities, holes or pockets. If allowed by the procedure, blotting or even a filtered air spray at less than 25 psi (172 kPa) may be used, but care should be used not to smear any indications that may begin to bleedout.

The drying method will depend on the type of developer used and the test object. The drying time should be the minimum required so that development and evaluation can be performed as quickly as possible. When hydro-air nozzles are used, the air pressure should not exceed 25 psi (172 kPa).

Manual Wipe (Method C)

After the required penetrant dwell time, the test object is first wiped with a clean, dry, lint free cloth or absorbent towel. The towel should be white, or a color that contrasts with the penetrant. The removal of the bulk excess penetrant works best if clean sections of cloth are used for each wipe.

The removal process should start from the most difficult areas to clean, such as areas of acceptable weld undercut or ripples. After the bulk of the excess penetrant is removed with dry cloths, the remaining penetrant is removed with dampened cloths. In no case should the cloths be saturated with cleaner. If any cleaner drips or can be squeezed from the cloth, it is too wet and very shallow indications may be removed.

When the surface of the test object is visually free from penetrant and the wipe cloth is coming out relatively clean, the wiping process is complete. Then the technician watches the solvent or water evaporate until the surface is dry, and then the required type of developer may be applied. Some procedures allow a dryer oven with recirculating air, typically not warmer than 71 °C (160 °F).

Developer Application and Drying

There are four basic types of developers.

1. Dry powder (fine powder form).
2. Water soluble (liquid dip tanks).
3. Water suspendible (liquid dip tanks).
4. Nonaqueous (aerosol spray can).

All developers must be applied so that a thin uniform coating covers the entire test surface. Application of an excessive thickness can obscure, cover or extinguish both visible and fluorescent indications. The classifications and proper application are as follows.

Dry Powder

Dry developer is loose, fluffy powder used with fluorescent penetrants. After removal of excess penetrant and/or emulsification and drying, dry powder developer is applied to the test surface for the purpose of absorbing penetrant from discontinuities and enhancing the resultant penetrant indications.

Dry developer may be applied in a dust cloud chamber activated by an air blast. The developer can be applied with an electrostatic sprayer, or the technician can dip the test object into the dry powder. Excess powder is shaken off of the test object.

Of the different developers available, dry developer is the most adaptable to rough surfaces and automatic processing. It is also the easiest to handle, apply and remove. Dry powder is not corrosive, gives off no vapors and leaves no residue or film to affect the next processing step. Dry powder sensitivity is slightly lower than water soluble for smooth test objects, but it is more sensitive for rough test objects or threads. Dry developer should not be used with visible dye penetrant because of the poor contrast provided by the thin coating of very fine powder.

Water Soluble

Water soluble powder crystals are mixed with water and dissolve in the solution, following the manufacturer's recommendations. The developer solution mixture concentration is verified by checking the specific gravity of the liquid with a hydrometer after mixing. Water soluble developer concentration will change over time due to evaporation losses. Therefore, the developer concentration should be checked on a weekly basis. The application occurs immediately following excess penetrant removal and/or emulsification from the test object and after draining or shaking off excess water, but before drying time.

Wet developer is applied by dipping (immersion), flow or spray techniques. Immersion in a prepared tank of developer is the most common application method. With immersion testing, the test object is immersed only long enough to coat all surfaces. It should then be removed immediately, and all excess from recesses or trapped areas should be drained to prevent pooling of developer, which can obscure indications. Wet developer is applied to form a smooth, even coating.

Particular care should be taken to avoid concentrations of developer in dished or hollowed areas of the test object. Such concentrations of developer may mask penetrant indications and are to be avoided. This developer in not normally used for critical applications on complex geometry test objects.

Water Suspendible

The water suspendible application is a suspension of white powder mixed with water, following the manufacturer's recommendations. Water suspendible developers require constant mild agitation (or thorough agitation before and during use) to keep the powder particles in suspension. Water soluble developer does not require agitation.

The water suspendible mixture concentration is verified by checking the specific gravity with a hydrometer after mixing. Water soluble developer concentration will change over time due to evaporation losses. Therefore, the developer concentration should be checked on a weekly basis. The application immediately follows excess penetrant removal from the test object and after draining or shaking off excess water, but before drying time.

The wet developer is applied by dipping (immersion), flow or spray techniques. Immersion in a prepared tank of developer is the most common application method. The test object should be immersed only long enough to coat all surfaces. It should then be removed immediately, and all excess from recesses or trapped areas should be drained to prevent pooling of developer, which can obscure indications.

Wet developer is applied to form a smooth, even coating. Particular care should be taken to avoid concentrations of developer in dished or hollowed areas of the test object. Such concentrations of developer may mask penetrant indications and should be avoided.

The fact that constant agitation is required to maintain the proper mixture concentration is a disadvantage of water suspendible developers. Care should be taken not to generate foam during agitation because foam will cause uneven surface coating. This developer in not normally used for critical applications on complex geometry test objects.

Nonaqueous

Nonaqueous wet developer is a powder suspended in a volatile liquid solvent. The volatile liquid gives the advantage of decreasing the viscosity and increasing the liquid bulk in the cavity. This action forces the penetrant to the surface where it assist the dispersion of the penetrant away from the discontinuity. The evaporation of the solvent tends to pull penetrant into the developer.

Nonaqueous wet developer is the most sensitive developer. The application is by spraying, and the volatile liquid evaporates rapidly so that no delayed drying operation is required. The spray or aerosol container must be thoroughly agitated just before spraying, and spraying should be done sparingly so that a thin coating covers the entire test area. The sheen of the metal should barely be covered. Several very light coats are usually preferable, rather than attempting to cover the test area with one spray. A light check spray should be performed away from the test surface to ensure the spray nozzle is clean and free of obstructions.

Development Time

Proper development time is very important in the interpretation and evaluation of indications. To interpret an indication is to decide if it is false, relevant or nonrelevant. To evaluate an indication means to decide if it is acceptable, requires rework or causes the test object to be rejected.

Development time is the time allowed between the application of developer and the viewing of indications. If insufficient time is allowed, indications may not have time to develop. Too much time will cause indications to become blurred or distorted.

The required interpretation time depends on the developer, and is stated in the procedure or specification. With a nonaqueous wet developer, the normal time until evaluation is 10 min to 1 h. With an aqueous developer, the normal time until evaluation is 10 min to 2 h. With a dry developer, the normal time until evaluation is 10 min to 4 h.

INTERPRETATION AND EVALUATION

The terms interpretation and evaluation are often confused. Actually, the terms refer to two entirely different steps in the testing process. The success and reliability of the interpretation and evaluation of liquid penetrant test indications is dependent on the thoroughness, training and dedication of the technician. The liquid penetrant test is not a method by which a test object is processed through a machine that separates the good test object from the bad. Testing personnel are required to carefully process each test object, interpret indications and then evaluate the indication to the acceptance standard referenced or given in the procedure they are performing.

Interpretation

To interpret an indication is to decide what condition caused it. The technician must first determine if the indication is a false, relevant or nonrelevant indication. One method of determining if an indication is relevant is to dampen a cotton swab with solvent and gently wipe the indication off. If the indication reappears, it is a relevant indication. This technique is known as the *bleed back evaluation technique*.

If the indication is not a relevant indication, the technician must determine the source of the invalid indication and correct the problem. Test objects with nonrelevant indications must be recleaned and reprocessed.

Valid indications may be relevant or nonrelevant. Relevant indications are caused by a surface discontinuity in the test object. Nonrelevant indications are caused by a design feature, such as a press fit or a rivet. When the technician determines that the indication is relevant, it must then be evaluated to the acceptance criteria.

Evaluation

To evaluate an indication means to decide if it is acceptable, requires rework or causes the test object to be rejected. Most industry acceptance standards are written with clear specifications. For some applications of liquid penetrant testing, a technician may only be told that no cracks or no linear indications are allowed.

When inspecting for material discontinuities, the technician determines if the indication is a crack, seam, lap, porosity, lack of fusion or hot tear. This determination depends on the knowledge of the test object's history, and the company policy for evaluation. In some cases, a metallurgist may be asked to determine the cause of the discontinuity.

The next step is to measure each indication and fill out a detailed report, including a sketch or picture to submit for further engineering evaluation. In many cases, a Level II technician evaluates and either accepts or rejects the test object. It is very important to mark and separate rejected test objects.

INDICATIONS

A liquid penetrant indication is caused by a discontinuity on the surface or by penetrant remaining on the surface from a false or nonrelevant cause. However, proper interpretation requires familiarity with the manufacturing processes, the types of discontinuities likely to occur and their appearance. It requires a knowledge of how different materials fail and where failure occurs. It requires the careful control of dwell and development time. The effects of these variables are discussed in the following sections.

False Indications

The most common source of false indications is poor cleaning of test objects and poor removal of excess penetrant. Ultraviolet radiation is very important during the washing process when using fluorescent penetrant. The technician determines whether a good rinse is obtained or whether patches of fluorescence remain on the test object. Adequate lighting at a fluorescent rinse or cleaning station is usually considered 100 lx (10 ftc) of white light and 100 $\mu W/cm^2$ of ultraviolet radiation at the test object surface.

For solvent removable penetrants, the removal process is controlled by the technician. This allows for a slower and more thorough test. To guard against confusion resulting from fluorescent or color spots other than relevant indications, care is taken so that no outside contamination occurs. Typical sources of contamination include the following.

1. Penetrant on the hands of the technician.
2. Contamination of wet or dry developer.
3. Penetrant rubbing off of an indication on one test object to a clean portion of the surface of another test object.
4. Penetrant spots on the testing table.

To avoid contamination, process tanks and testing areas are kept clean. Only lint free wiping cloths or rags are used, and test objects are kept free of fingerprints and tool marks. Figure 2.1 illustrates common types of false indications, caused by handling or cleaning processes.

Figure 2.1: Typical false indication.

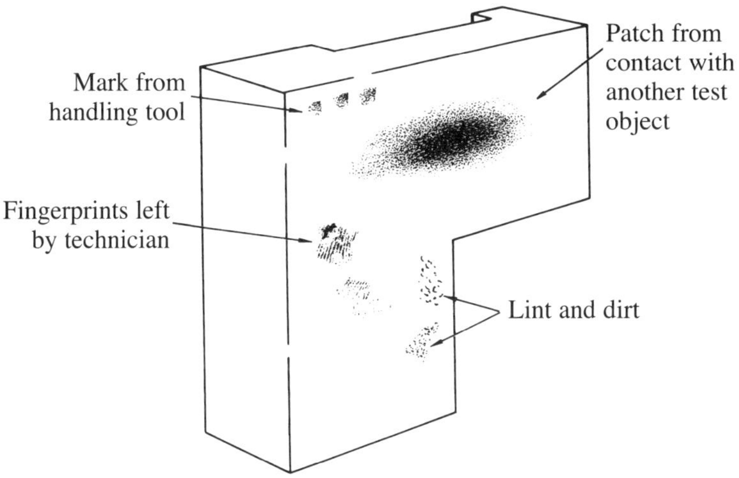

Nonrelevant Indications

Nonrelevant indications are caused by features in the test object that are there by design, but are in no way a relevant discontinuity. Most nonrelevant indications are easy to recognize because they are related directly to some feature of the assembly that accounts for their presence.

Nonrelevant indications include those that appear on test objects that are press fitted, riveted or spot welded together, and those caused by surface roughness.

Any nonrelevant indication that interferes or could mask a relevant indication must be further evaluated and sometimes retested. If a test object is too rough to perform a proper liquid penetrant test, or has a condition like weld undercut that has been accepted visually, the test object or weld may have to be returned for further preparation for liquid penetrant testing.

Relevant Indications

Relevant indications are those caused by a material discontinuity. The interpretation of an indication as relevant is a matter of observing the indication, eliminating the possibility of it being a false indication and then further determining that it is not a nonrelevant indication. Any relevant indication immediately becomes subject to evaluation of its cause (type of discontinuity).

Some procedures allow measuring the visible discontinuity size after wiping off the developer. This should be done very carefully, with magnification and reapplication of developer to compare the indication size measured visually to the penetrant indication. For example, a small diameter, deep, visible porosity may bleedout as a very large indication after the minimum dwell time. This fact should be reported.

Categories of Relevant Indications

Discontinuity indications vary widely in appearance, but for each indication two interpretive questions must be answered.

1. What type of discontinuity caused the indication?
2. What is the extent of the discontinuity as evidenced by the extent of the indication?

The answers to these interpretive questions are obtained by observing the indication and identifying the discontinuity by evaluating the characteristic appearance of the indication.

Each indication may also require an answer to the evaluation question: What effect will the indicated discontinuity have on the service life of the test object?

The answer to the evaluation question is based on certain knowledge of the seriousness of the discontinuity and complete understanding of the test object.

Figure 2.2 illustrates typical relevant indications. Relevant indications fall into five categories.

1. Continuous line indication.
2. Intermittent or broken line indication.
3. Rounded indication.
4. Faint indication.
5. Gross indication.

Figure 2.2: Typical relevant indications.

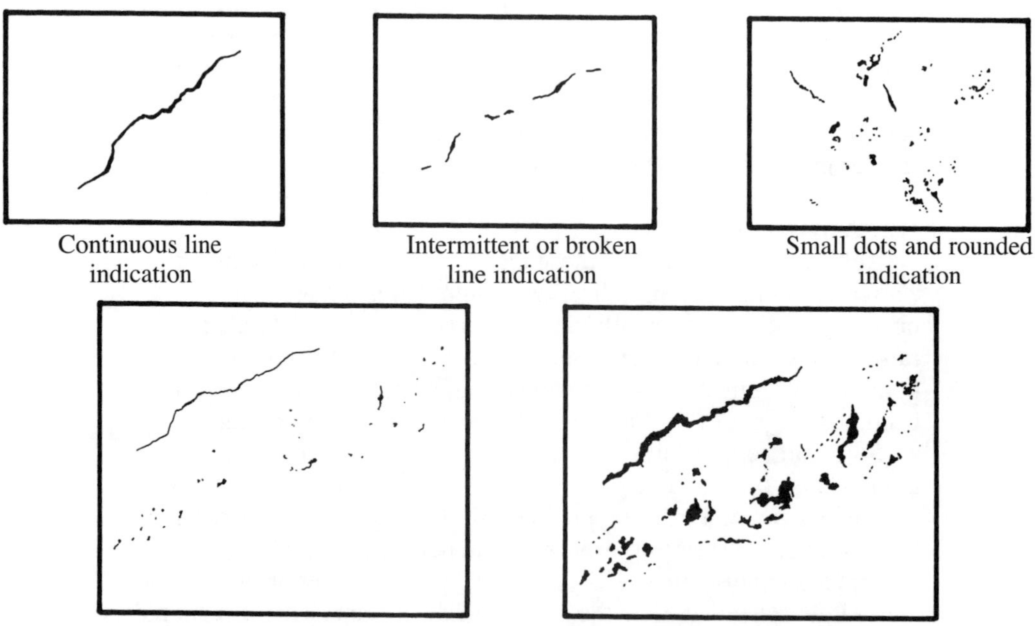

Continuous line indication — Intermittent or broken line indication — Small dots and rounded indication

Faint indication — Gross indication

Continuous Linear

Continuous linear indications are caused by cracks, seams, cold shuts, forging laps, scratches or die marks. Linear indications are those that are three times greater in length than width. Cracks usually are linear and irregular, rather than straight. Seams appear as straight or spiral lines. Cold shuts appear as smooth, narrow, regular solidification lines. Forging laps appear as smooth lines. Scratches and die marks appear in a variety of linear patterns, but are recognizable when all penetrant traces are removed since the bottom of the discontinuity is usually visible. Scratches, die marks and other surface discontinuities are stress risers, and some test objects require evaluation of these conditions.

Intermittent Linear

The same discontinuities that cause continuous linear indications may, under different circumstances, cause intermittent linear indications. When a test object is worked by grinding, peening, forging or machining, portions of the discontinuities on the surface of the test object may be closed or removed by the metal working process. When this occurs, the discontinuities will appear as intermittent lines. Machining operations may also open linear indication that lie below the surface, exposing only the tops of continuous indications and thus making them appear as intermittent indications. Some acceptance standards consider separation of indications and require evaluation and reporting of the total length to the extremities of the indications. Most specifications require some type of etch to be used if metal smearing has occurred.

Round

Round indications are usually caused by porosity. The porosity may be the result of gas holes, weld porosity or the generally porous makeup of the test object. Round indications are those that are less than three times greater in length than width. Deep cracks (especially weld crater cracks) may also appear as larger round indications after they bleedout because they trap a large amount of penetrant that spreads when the developer is applied.

Any large, round liquid penetrant indication may appear as an elongated porosity on a radiograph, and may have some length if tested by ultrasonic testing. After evaluation, a technician may wipe away a large, round indication and reapply developer, then watch the formation during the bleedout to obtain more information about the nature of the discontinuity.

Diffused or Weak

Diffused or weak indications are particularly difficult to interpret. Weak indications appearing over a large area are always suspect. When they appear, the test object should be thoroughly cleaned and retested. If the indications were not caused by inadequate cleaning, the test object may require more preparation.

Discontinuity Depth Determination

When evaluation requires more accurate knowledge of the relative depth of a discontinuity, it may be obtained by removing the surface indication and reapplying developer. If the subsequent amount and rate of penetrant bleedout is fast, large and bright, then it is a deep discontinuity. This is why many acceptance standards reject large, rounded indications. There is no way to estimate a dimension for the depth of a discontinuity.

Typical Indications

Figures 2.3 to 2.26 illustrate typical indications found in liquid penetrant testing.

Figure 2.3: Fluorescent streaks and patches caused by incomplete wash.

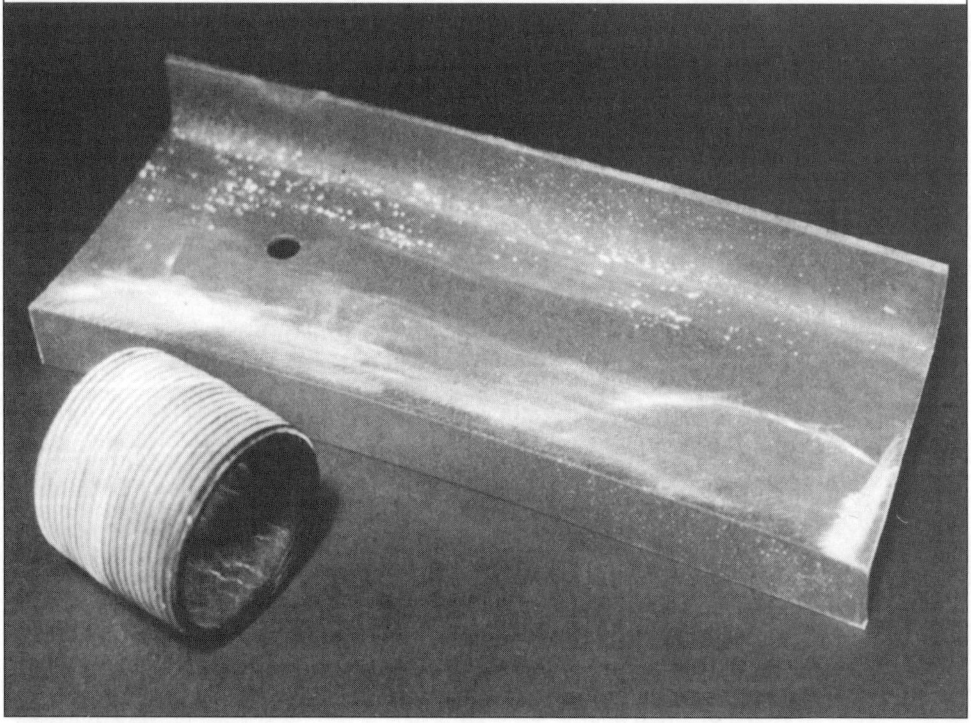

Figure 2.4: Nonrelevant fluorescent indication caused by rough surfaces and poor wash.

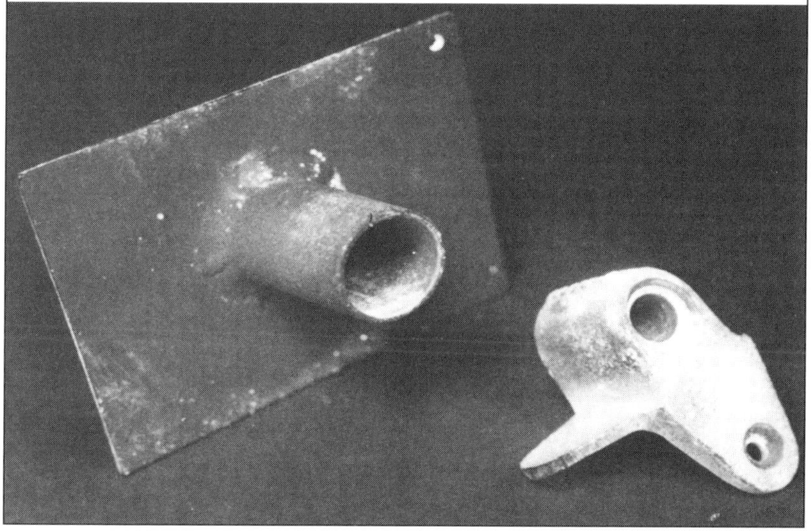

Figure 2.5: Visible dye penetrant indications of surface bursts in steel forging.

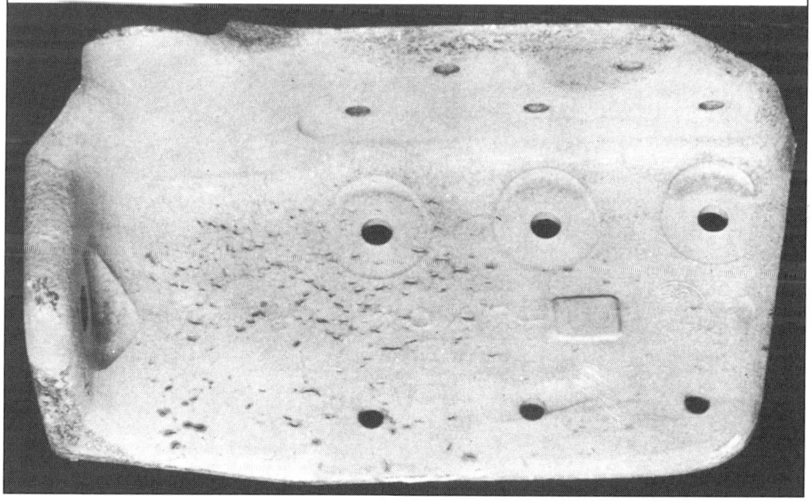

Figure 2.6: Visible dye penetrant indication of forging lap.

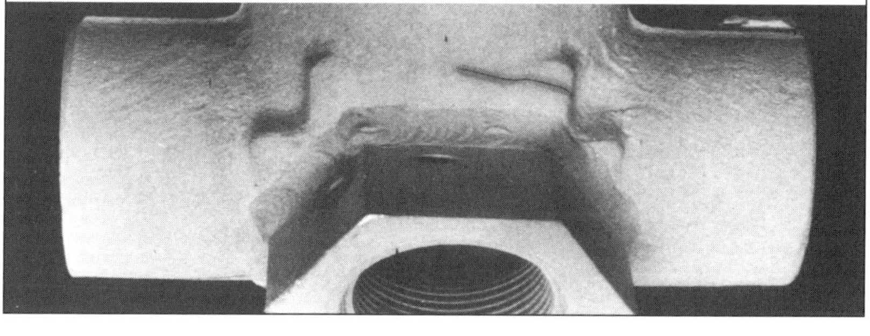

Figure 2.7: Visible penetrant indication of lamination.

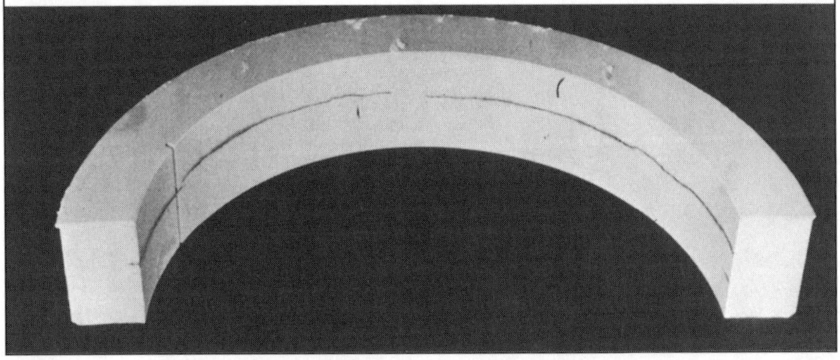

Figure 2.8: Visible penetrant indications of weldment cracks.

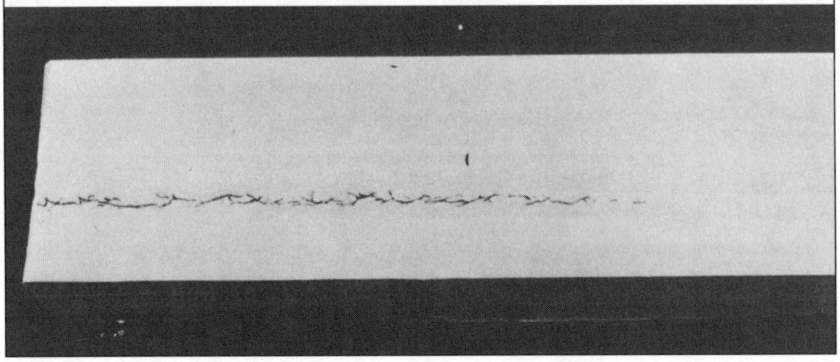

Figure 2.9: Visible penetrant indications of cracks in aluminum forging.

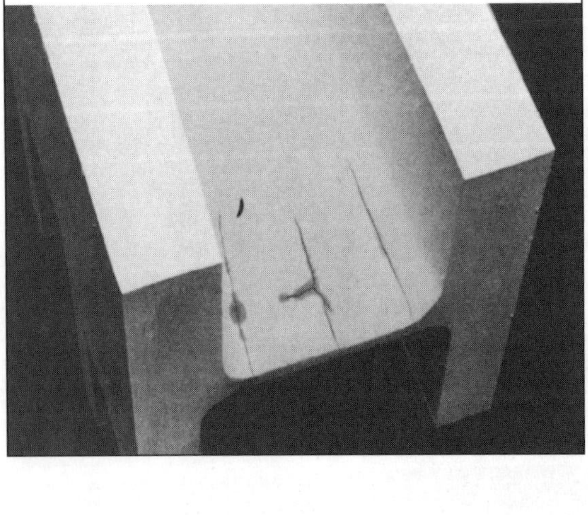

Figure 2.10: Visible penetrant indications of unhealed porosity in aluminum plate.

Figure 2.11: Fluorescent indications of porosity in machined aluminum plate.

Figure 2.12: Fluorescent indications of blow holes and random porosity in aluminum casting.

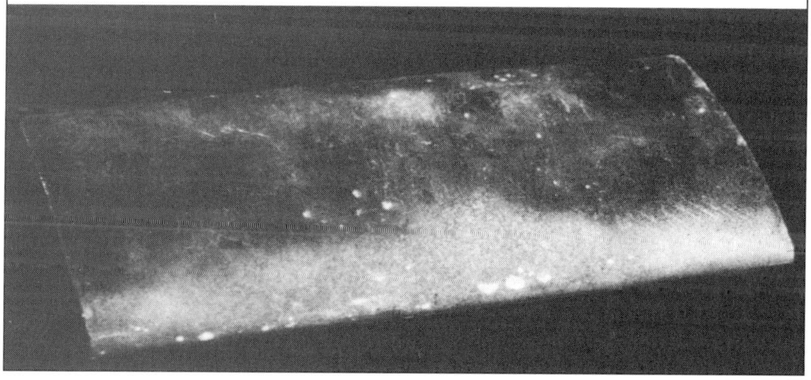

Figure 2.13: Fluorescent indications of tears in stretch formed aluminum angle.

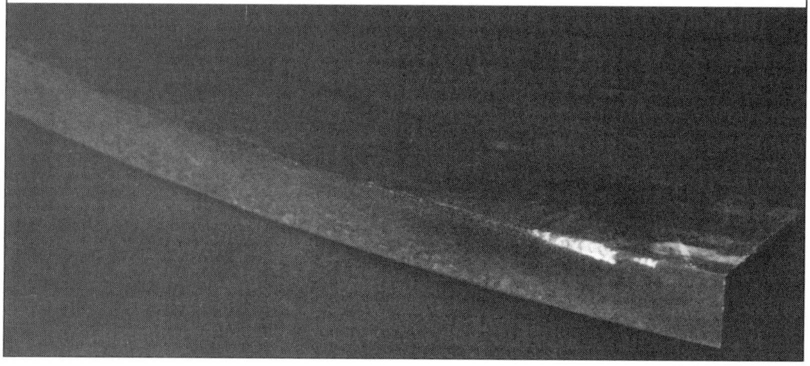

Figure 2.14: Fluorescent indications of radius cracks in machined steel fitting.

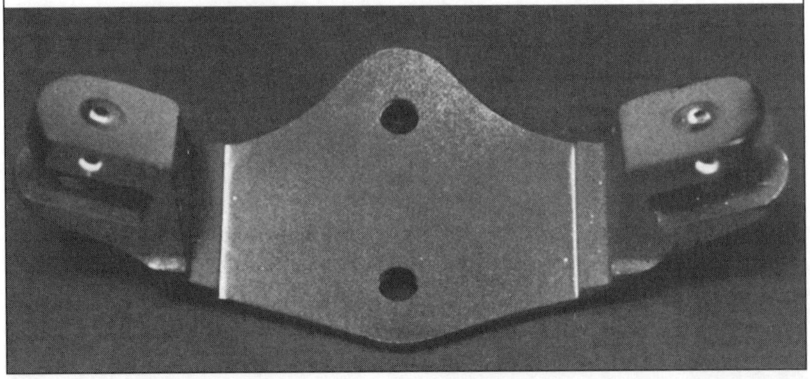

Figure 2.15: Fluorescent indications of thermal cracks, porosity and seams in unglazed ceramic coil form.

Figure 2.16: Fluorescent indications of cracks and pores in stainless steel weld.

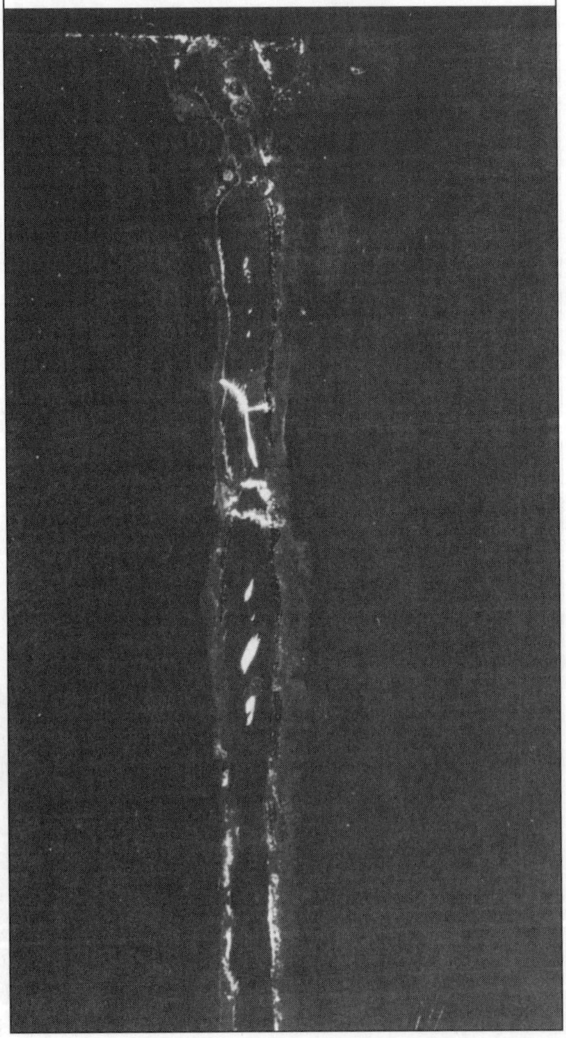

Figure 2.17: Fluorescent indications of cracks produced in hard fired unglazed ceramic rod.

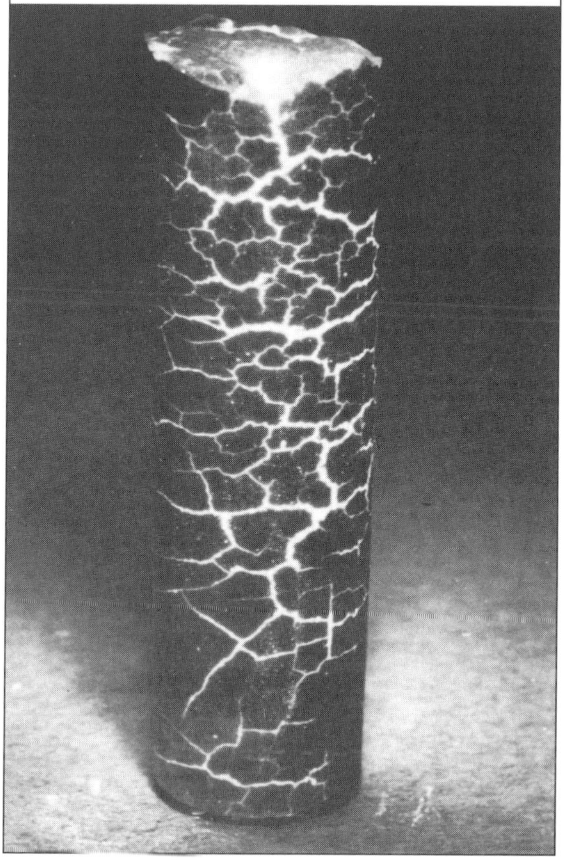

Figure 2.18: Fluorescent indications of cracks in unfinished coupling castings.

Figure 2.19: Fluorescent indications of surface porosity in magnesium sand casting.

Figure 2.20: Fluorescent liquid penetrant indications of porosity and shrinkage cracking in cast light alloy aircraft fitting.

Figure 2.21: Fluorescent liquid penetrant indications of crack in stainless steel weldment.

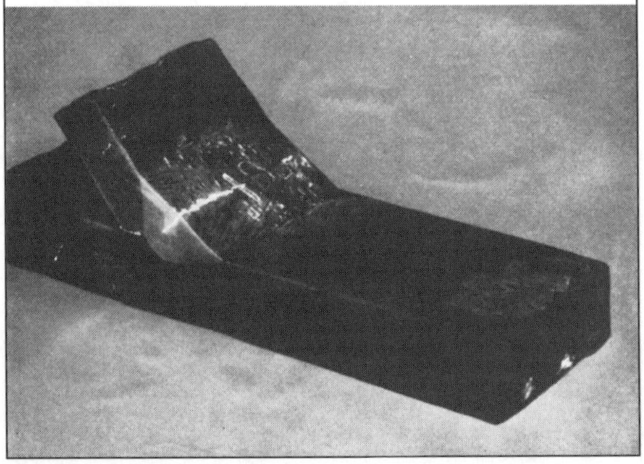

Figure 2.22: Fatigue crack in magnesium weld.

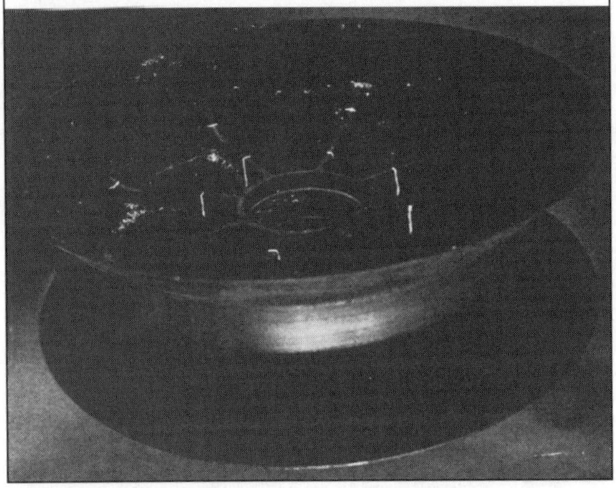

Figure 2.23: Forging laps in aluminum forging.

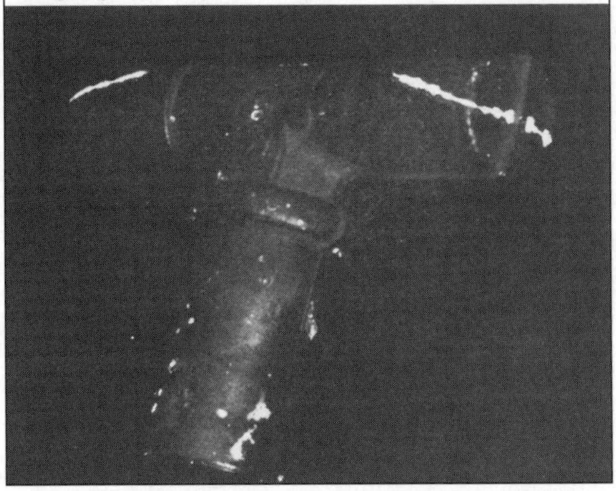

Figure 2.24: Typical quench cracks in 4340 steel are revealed by high performance water washable liquid penetrant with nonaqueous wet developer.

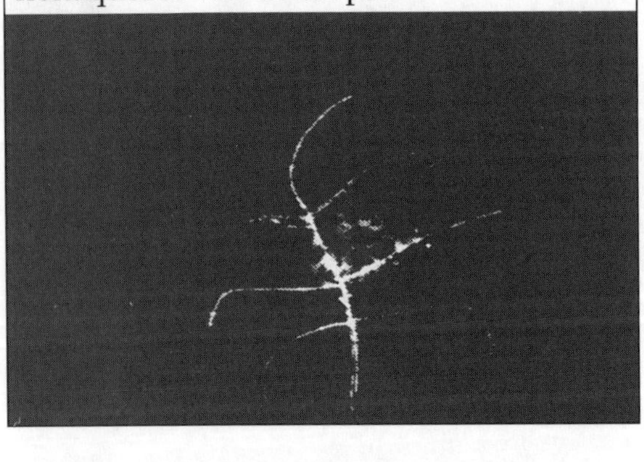

Figure 2.25: Crack indications on connection outside diameter after development time.

Figure 2.26: Stress corrosion cracking in nuclear grade type 304 stainless steel piping: (a) white light photograph of typical pipe weld segment; (b) fluorescent liquid penetrant indications of cracks angled to and parallel to weld center line; (c) closeup of photograph of long, parallel fluorescent crack indication in Figure 2.26b; (d) closeup of photograph of short, angled crack indications in Figure 2.26b; (e) photograph of sample, broken through stress corrosion crack; and (f) cross section of typical stress corrosion crack in pipe wall.

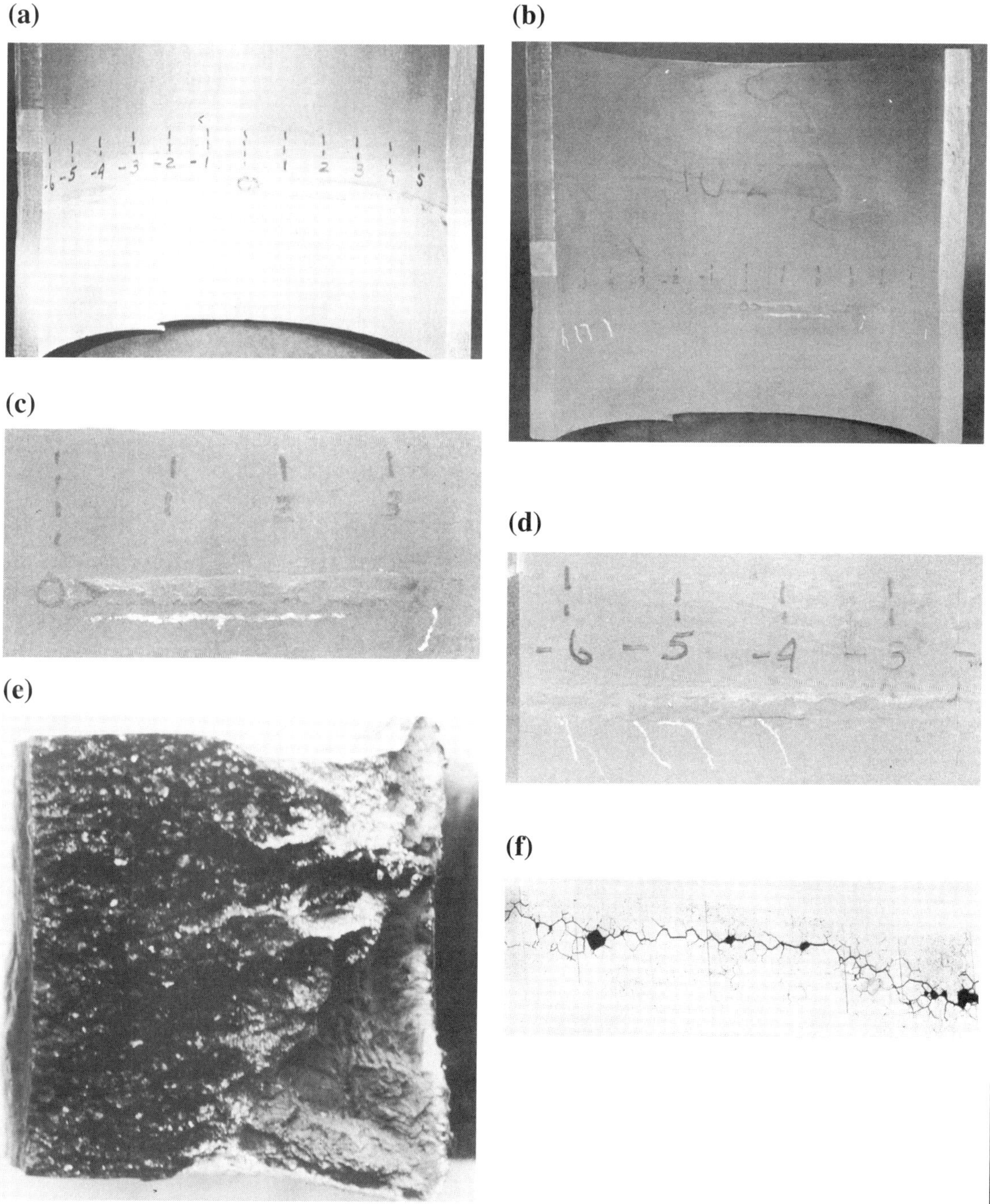

Postcleaning

If all traces of penetrant testing materials are not removed after the test, they may have a harmful effect when the test object is placed in service. The cleaning processes typically used with liquid penetrant testing are discussed in the following chapters. In some cases, the preparation for the next production process, such as plating or painting, may be sufficient. Components shall be cleaned after examination to remove developers and other examination material residues if these are detrimental to subsequent operations or to the test object's intended function.

Postcleaning Materials

Following the test, the residue of penetrant materials may have to be removed. Postcleaning is particularly important when test objects are destined for use in an oxygen environment. In fact, oxygen compatible penetrant materials must be used to test these objects. Though many test objects will receive further processing, the cleanliness of any test object after completion of a penetrant test is the responsibility of test personnel.

Except for liquid oxygen compatibility and the chlorine free requirement in the precleaning and postcleaning of nickel alloys, certain stainless steels and titanium, no special materials are required for postcleaning unless required by specification or company procedures.

Chapter 3

Liquid Penetrant Testing Methods

Standard Methods

There are several organizations that issue standards for liquid penetrant testing. There are minor differences in these standards, which are easily incorporated into a single liquid penetrant procedure that covers the two types and four methods of liquid penetrant testing (see Table 1.1 in Chapter 1). Each company performing liquid penetrant testing is required to have its procedures approved for the basic standard requirements.

Method Characteristics

Fluorescent or Visible

The choice to use Type I (fluorescent) or Type II (visible) penetrant is sometimes based solely on the availability of a dark area and power for the ultraviolet radiation, but is more often based on the necessity of the specified sensitivity levels available only with Type I (fluorescent) penetrant processing. All the methods require the same thorough precleaning and drying before penetrant application.

The greatest difference between Type I (fluorescent) and Type II (visible) systems is the sensitivity rating system for Type I (fluorescent). Type I (fluorescent) systems are generally more sensitive than Type II (visible) systems. Because of this greater sensitivity, the use of Type I systems is required on most critical aerospace components. Type II (visible) penetrant should never be used before Type I (fluorescent) penetrant, because any red penetrant remaining in discontinuities will dilute and the indication will not appear under ultraviolet radiation.

Solvent or Water Removable

The portable solvent or manual water removable penetrant kits allow field tests at construction sites, aboard ships or on large assemblies in fabrication shops. The materials are in aerosol spray cans. This allows one technician to carry all the equipment and accomplish the test. The Type II (visible), Method C (solvent removable) test is the most common portable method, but in some cases water removable penetrant techniques can be used. Portable fluorescent kits are also available. Aerosol cans do not require daily checks or contamination checks.

Inline Penetrant Systems

Inline penetrant systems consist of bulk penetrant materials in tanks. Type I (fluorescent) is preferred because the indications are easier to see. The size and design of the penetrant line depends on the type and size of the test objects. The tanks can be very small in a separate room, or built so the entire assembly line passes through the penetrant process. The manual water rinse is the most common penetrant removal method. Emulsification dip tanks or sprayers may be used if required for certain test objects. Inline penetrant systems are used to test a large volume of manufactured or cast test objects, but are also commonly used for overhaul and maintenance shops.

Method A (water washable) is the most common, fastest and simplest penetrant method. Method A (water washable) penetrant is normally sensitivity Level 2 or 3, but penetrant with sensitivity Levels 1/2, 1, 2, 3 and 4 are available. The test object and specification requirement will determine which sensitivity level should be used. The clean, dry test objects are typically dipped briefly in the tank of penetrant, then moved to a drain station that allows the excess penetrant to drain back into the penetrant tank during the dwell time.

After the required dwell time, the test objects are taken to the rinse station, which has appropriate lighting that will be a darkened area of less than 100 lx (10 ftc) white light and greater than 1000 μW/cm^2 of near ultraviolet radiation (UV-A). The spray nozzles are designed to put out a coarse droplet spray. The water is less than 40 psi (275 kPa), and the temperature is regulated to remain between 10 to 38 °C (50 and 100 °F). The water is sprayed from a distance of about 30 cm (12 in.) minimum at an oblique angle. The background fluorescent lighting makes it easy to determine when the excess penetrant is removed. If nonaqueous or dry developer is used, the test object is then ready for the drying oven. If aqueous developer is used, the test object is then dipped in wet developer.

The postemulsified Methods B or D can also be added to a liquid penetrant line by adding the emulsifier dip tank(s) or the spray equipment. The postemulsified methods are normally used when very shallow discontinuities are critical, or when every possibility of eliminating over rinsing of test objects is a requirement. Critical objects, such as turbines or aircraft engine parts, often require high sensitivity postemulsification methods. Special restrictions and requirements may be specified.

Lipophilic Emulsification

Method B is the lipophilic (oil based) emulsification method. After the proper oil base penetrant is applied and the dwell time has elapsed, the test object is ready for emulsification. The lipophilic emulsifier is typically located in a dip tank, and is a contrasting color from the penetrant so that it forms a visible coat over the penetrant to ensure complete coverage.

Application is typically by dipping, but flowing may be used. No agitation of the test object is allowed. The lipophilic emulsifier acts by diffusing into the oil base penetrant and scraping the penetrant from the surface, making it water washable.

The emulsification time is determined by experimentation and depends on the features of the test object, the uniform dipping and draining of the test object and the viscosity of the emulsifier. This makes uniform dipping, draining and emulsification time very important, so this method is typically monitored in seconds. The maximum time allowed is 180 s for Type I systems and 30 s for Type II systems, and will be shorter for smoother test objects. After the proper emulsification, the water rinse is identical to the Method A rinse and developing will be the same as other methods.

Hydrophilic Emulsification

Method D is the hydrophilic (water based) emulsification method. The test object is ready for emulsification after the proper oil base penetrant is applied and the dwell time has elapsed. The hydrophilic emulsifier can also be used in tanks, or spray equipment can be modified so a mix of hydrophilic emulsifier and water can be used to remove the oil based penetrant.

Hydrophilic emulsifier is water based and is supplied in a concentrated form that is diluted in water to concentrations of 10 to 30% for dip applications, and 0.05 to 5% for spray applications. A water prerinse is required to help remove some of the bulk penetrant before dipping in the emulsifier. This helps prevent some of the penetrant contamination in the emulsifier tank.

Hydrophilic emulsifier acts on the penetrant from the surface by detergent action. The spray or agitation in the tank provides a scrubbing action. The emulsification time for dip tank applications is determined by experimentation and is normally 120 s maximum. The manufacturer will specify the proper concentration of emulsifier in water, which should be checked periodically with a refractometer.

Tanks of emulsifier will mix with a small amount of penetrant during the dipping process over a period of time. This is one reason for periodic checks to monitor the system materials and performance.

Chapter 4

Liquid Penetrant Testing Equipment

Liquid Penetrant Testing Units

Liquid penetrant testing units can be arranged so test objects are moved manually, mechanically assisted, semiautomatic or fully automatic. This choice depends on budget, timing and the long term system performance. The size of the testing unit is largely dependent on the size and types of the test objects. The layout of the equipment may be *U* shaped, *L* shaped or straight line, and is determined by the facilities available, the production rate and the required ease of handling.

Stations

Depending on the type of penetrant and processing used, the liquid penetrant testing units require stations as shown in Figures 4.1 through 4.5. In a typical testing facility for a postemulsification process, the following stations are required.

1. Precleaning station (usually remote from penetrant test station).
2. Penetrant station (tank, tunnel or application booth).
3. Drain station (used with penetrant tank).
4. Prerinse station.
5. Emulsifier station (tank, tunnel or application booth).
6. Rinse station (sink with lighting).
7. Developer station (tank, tunnel or application booth).
8. Dryer station (usually an oven).
9. Inspection station (enclosed booth or table with proper lighting).
10. Postcleaning station (usually remote from penetrant test station).

Auxiliary Equipment

Auxiliary equipment is defined as the equipment located at penetrant test stations (other than cleaning stations) required to perform penetrant testing. In some instances, the auxiliary equipment may be built in at one or more of the test stations.

Figure 4.1: Typical small test equipment using fluorescent postemulsifiable penetrant and dry developer.

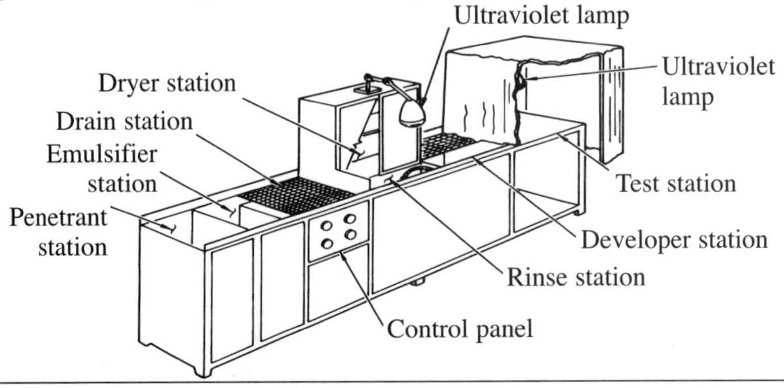

Figure 4.2: Typical medium test equipment using fluorescent water washable penetrant and wet developer.

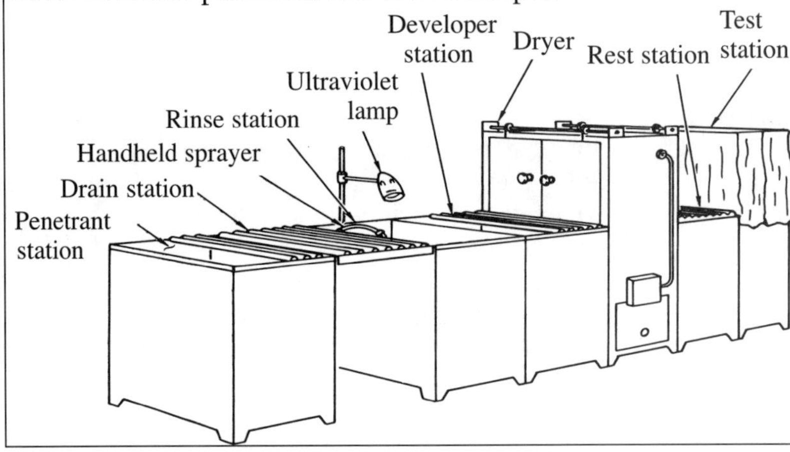

Figure 4.3: Typical large test equipment using fluorescent water washable penetrant and wet developer.

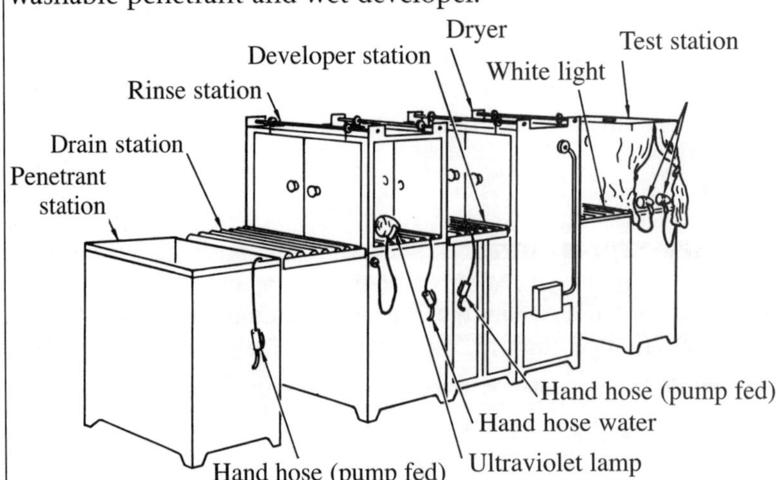

Figure 4.4: Modular construction of multiple component penetrant testing system permits straight line, *L* or *U* shaped arrangements: (a) liquid penetrant tank with cover; (b) drain and rest station; (c) wash station; (d) drying oven; (e) dry developer station; (f) testing station; (g,h) multiple component inline liquid penetrant testing system consisting of assembled units.

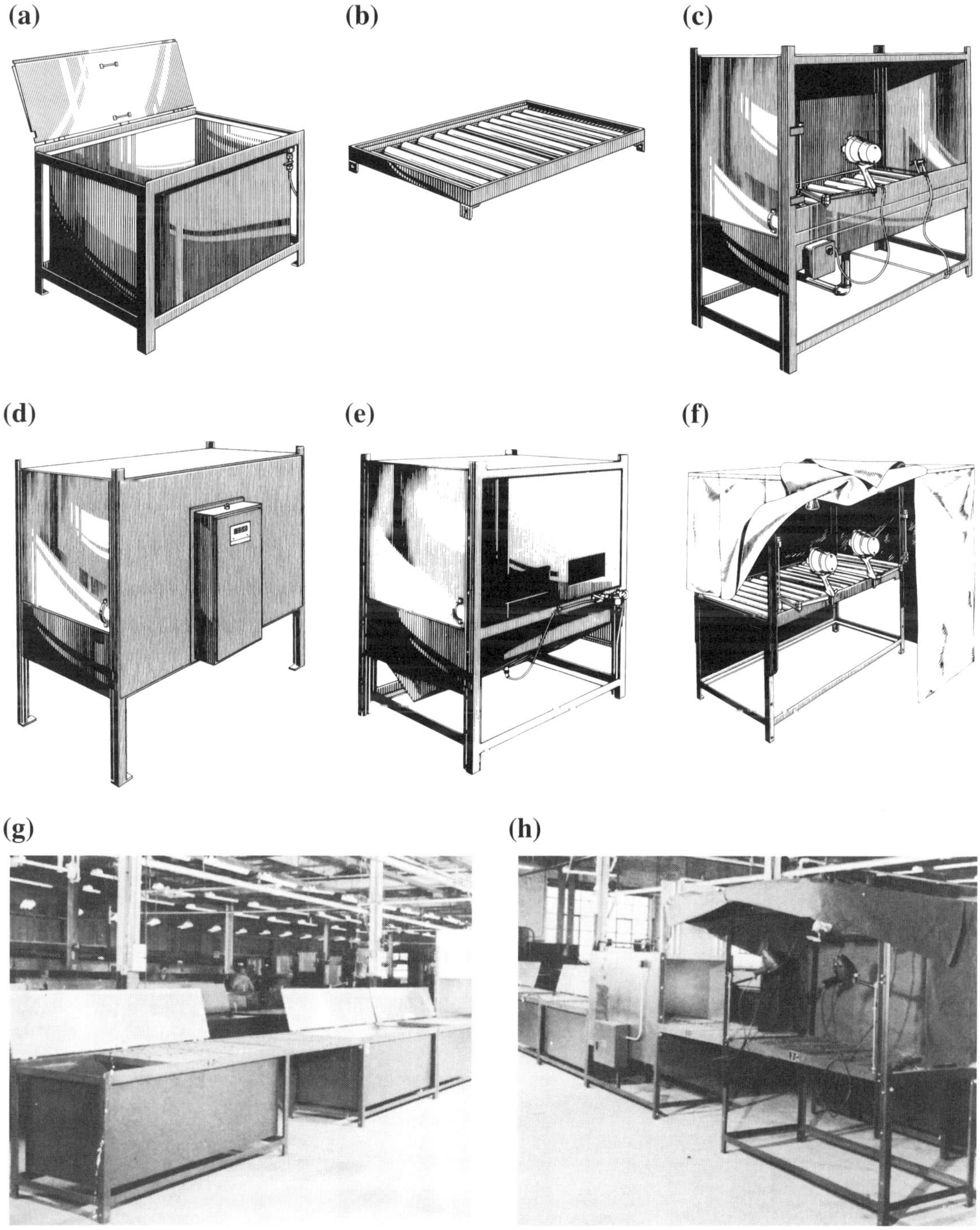

Figure 4.5: Small self contained water washable fluorescent liquid penetrant testing unit.

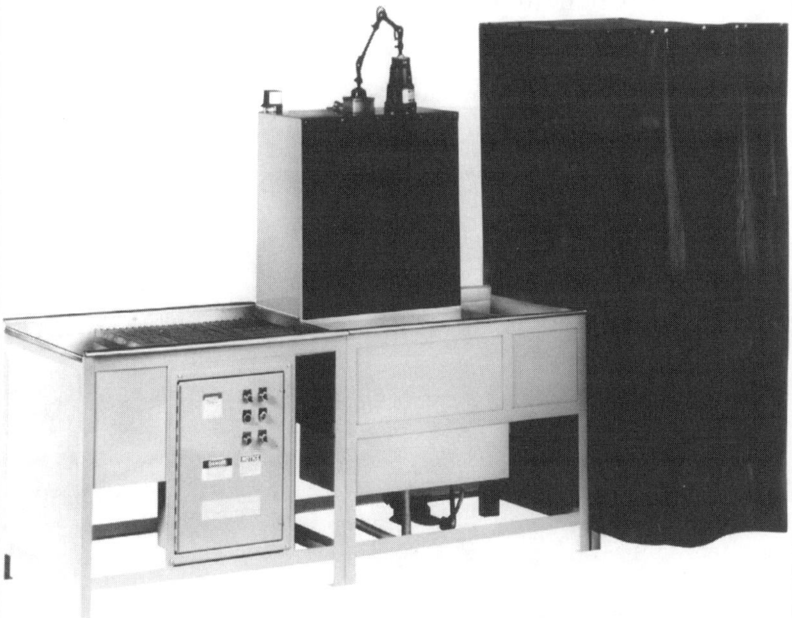

Pumps

Various pumps installed at the penetrant, emulsifier, rinse and developer stations are used to agitate the solutions, to pump drain off material into the proper tank for reuse or filtration for disposal and to power handheld sprayers and applicators.

Sprayers and Applicators

Sprayers and applicators are frequently used at the penetrant, emulsifier, rinse and developer stations. They decrease test time by permitting rapid and even application of penetrant materials. Both conventional and electrostatic sprayers are used for penetrants, dry developers and suspendible developers. An electrostatic spray system is shown in Figure 4.6.

Figure 4.6: Application with an electrostatic spray system minimizes consumption of liquid penetrant material and helps ensure coverage of complete surface: (a) diagram of electrostatic spray system components; (b) technician applies fluorescent liquid penetrant with a handheld electrostatic spray gun that makes atomized particles attract to the test object; and (c) two reciprocating automatic electrostatic spray guns in conveyorized installation are used to apply liquid penetrant to one side of the test object while other guns (not illustrated) operate from the opposite side further down line.

(a)

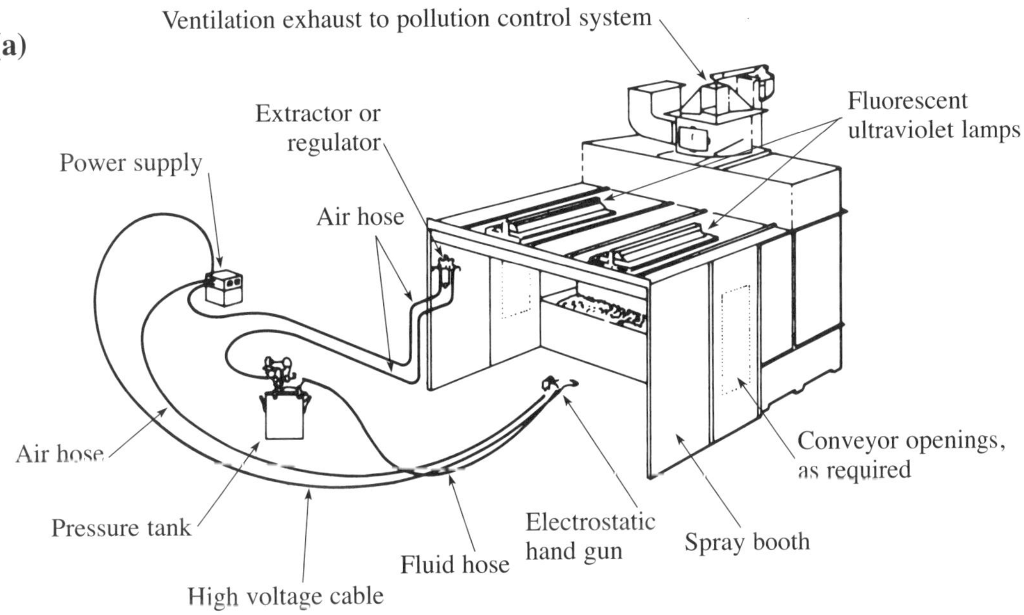

(b)

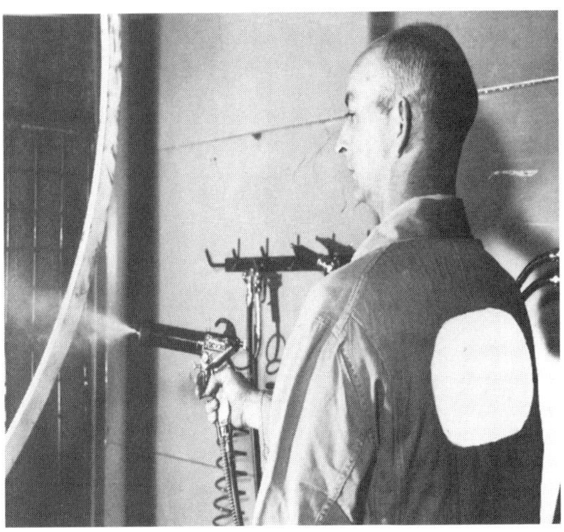

(c)

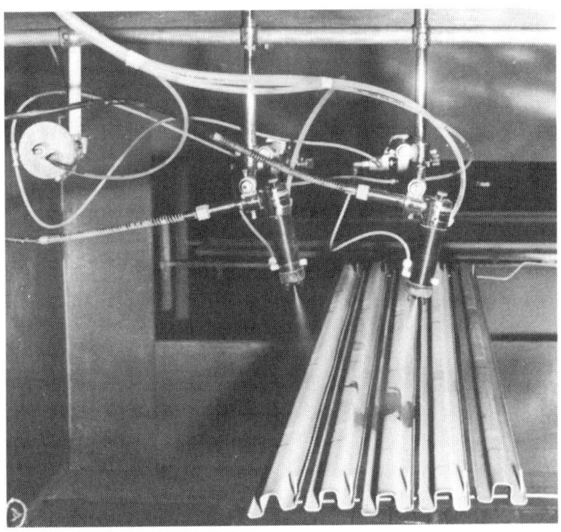

Lamps

White lamps, as well as ultraviolet lamps, are installed as required to ensure adequate and correct lighting at all stations. When fluorescent materials are used, ultraviolet lamps are installed at both the rinse and test stations.

Timers

One or more timers with alarms are used to control penetrant, emulsifier, developing and drying cycles.

Thermostats and Thermometers

Thermostats and thermometers are used to control the temperature of the drying oven and penetrant materials.

Exhaust Fans

Exhaust fans are used when testing is performed in closed areas. The fans facilitate removal of fumes and dust.

Refractometer

A refractometer is used to measure the concentration of hydrophilic emulsifier in water. A refractometer is shown in Figure 4.7.

Figure 4.7: Typical refractometer.

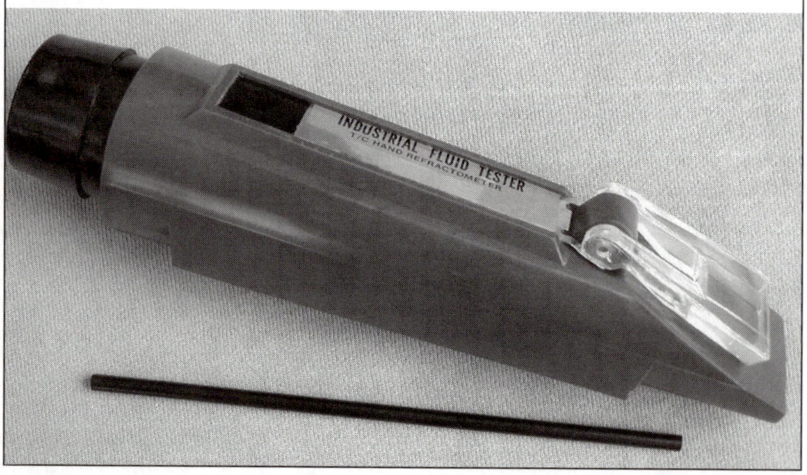

Hydrometer

The hydrometers used in liquid penetrant testing are floating instruments, as shown in Figure 4.8. They are used to measure the specific gravity of water based wet developers.

Figure 4.8: Typical hydrometer (shown horizontally, but used vertically).

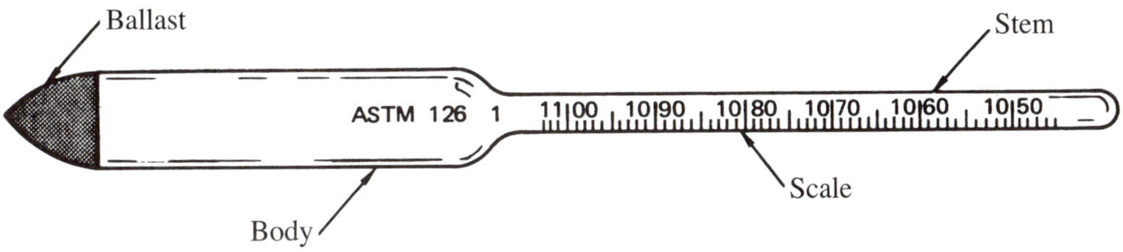

PORTABLE EQUIPMENT

It is possible to perform liquid penetrant tests without stationary equipment. When testing is required at a location remote from stationary equipment, or when only a small portion of a large test object requires testing, portable liquid penetrant kits may be used. Both fluorescent and visible dye penetrants are available in kits. The penetrant materials are usually dispensed from pressurized spray cans or the penetrant may be applied with a brush.

Visible Dye Penetrant Kit

The visible dye penetrant test kit is light in weight and contains all the materials necessary for a test, as shown in Figure 4.9. It consists of a container with at least the following items.

1. Solvent cleaner.
2. Penetrant remover.
3. Visible penetrant.
4. Application brushes or pads.
5. Nonaqueous wet developer.
6. Wiping cloths.

Figure 4.9: Typical visible portable kit.

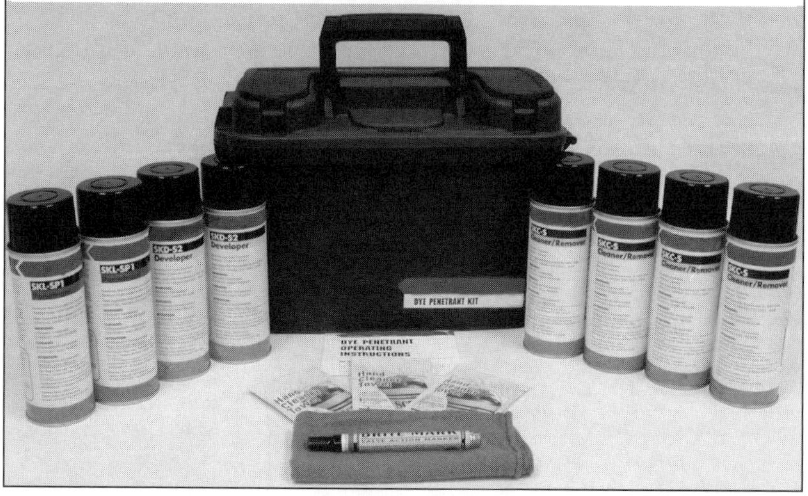

Fluorescent Penetrant Kit

The fluorescent penetrant kit combines portability with the high visibility associated with fluorescent materials. The kit holds all the essential materials required for the test, including an ultraviolet lamp, as shown in Figure 4.10. The fluorescent kit consists of a container with the following items.

1. Portable ultraviolet lamp.
2. Solvent cleaner.
3. Penetrant remover.
4. Fluorescent penetrant.
5. Application brushes or pads.
6. Developer, either nonaqueous wet or dry.
7. Wiping cloths.
8. Hood to provide darkened area for viewing indications.

Figure 4.10: Typical fluorescent penetrant portable kit.

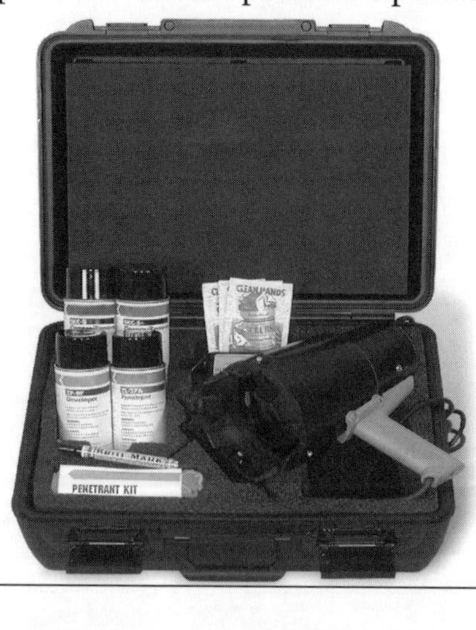

Ultraviolet Radiation

The electromagnetic spectrum including the ultraviolet radiation used for evaluation of fluorescent liquid penetrant indications is shown in Figure 4.11. Most common, handheld ultraviolet mercury vapor arc lamps with produce filtered light in the wavelength range of 320 to 440 nm. Ultraviolet mercury vapor arc lamps cause the fluorescent penetrant to fluoresce visible light in the green-yellow or red-orange wavelengths (between 520 and 580 nm), depending on the material being viewed. The equipment usually consists of a mercury arc bulb and a filter, as shown in Figure 4.12. The bulb and filter are contained in a reflector lamp unit.

For correct test results, the lamp should produce an intensity of at least 800 µW/cm^2 at the test surface, and most specifications require the output to be 1000 µW/cm^2 measured at 38 cm (15 in.), which should be checked on a daily or weekly basis. This equipment check is needed because the lamp loses its intensity over a period of usage and will have to be replaced before it burns out.

The lamp's deep red-purple filter is designed to pass only those wavelengths of light that will activate the fluorescent penetrant material, which is about 365 nm. It also helps filter out harmful ultraviolet radiation. Because dust, dirt and oil greatly reduce the intensity of the emitted light, the filter should be frequently cleaned.

When using mercury vapor lamps, the full intensity of the lamp is not attained until the mercury arc is sufficiently heated. At least 5 min warm up is required to reach the required arc temperature. Because switching the lamp on and off shortens bulb life, the lamp is usually left on during the entire test period. If the ultraviolet radiation is switched off, it will not respond immediately when turned on. It may take about 10 min to reach full intensity again.

Figure 4.11: Electromagnetic spectrum showing the narrow range of ultraviolet radiation used for fluorescent liquid penetrant testing.

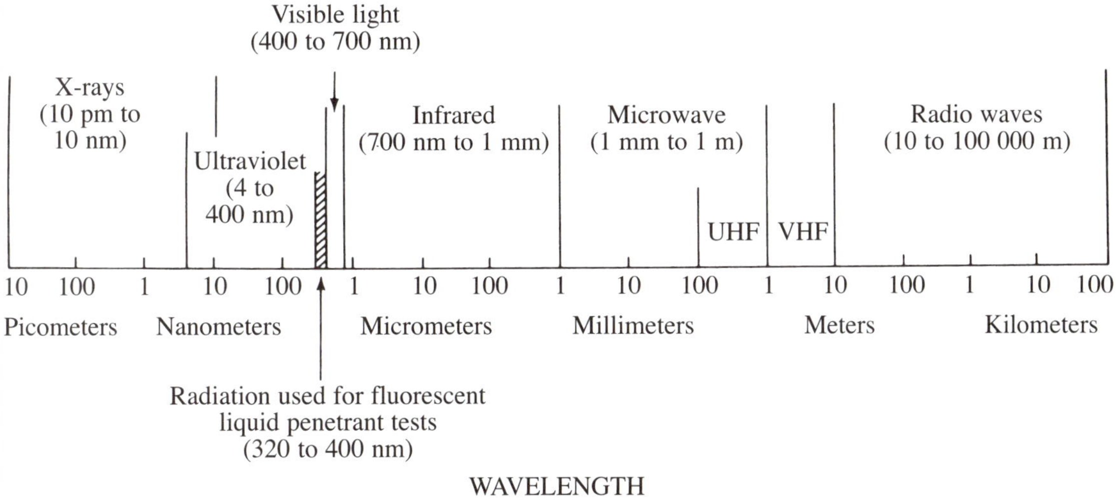

Figure 4.12: Ultraviolet radiation sources: (a) 100 W ultraviolet lamp with ballast transformer; (b) self ballasted ultraviolet lamp; (c) 12 V battery operated portable ultraviolet lamp with plastic housing; (d) portable, battery operated, tubular ultraviolet lamp; (e) 400 W ultraviolet flood lamp; and (f) low intensity, tubular ultraviolet lamp.

Fluorescent ultraviolet radiation tubes are also available that emit the 320 to 400 nm wavelengths, but focus is difficult and these bulbs are usually not adequate for evaluation stations. They are most commonly used for rinse station or auxiliary lighting.

Tubular Fluorescent Cold Discharge Sources

Electrically and mechanically, these standard fluorescent bulbs come in sizes from 2 to more than 60 W input. These cold discharge tubular lamps contain low pressure mercury vapor glow discharges. Their primary radiation is hard ultraviolet radiation of 253.7 nm wavelength. This is used to excite a special cerium activated calcium phosphate phosphor coated on the inside of the tube. This phosphor, when activated by ultraviolet radiation, emits ultraviolet radiation with a range of 320 to 440 nm, peaking at 360 nm. Because a significant amount of visible light is emitted along with the ultraviolet radiation, these bulbs are often made of a purple-red filter glass similar to that used over the high pressure arc lamps previously described. This greatly reduces the ultraviolet radiation emitted but still leaves what is often an excessive amount of visible blue light, considering the relatively low intensity of the ultraviolet radiation produced.

Testing in Subdued White Light

The improved liquid penetrant processing materials available now have made it possible to test in less than total darkness. This does not mean darkened inspection booths are no longer necessary. It does mean that with bright liquid penetrant indications and adequate ultraviolet radiation intensity, medium and large indications can be detected under conditions of fairly high ambient white light. This has made some fluorescent liquid penetrant testing possible on aircraft outdoors on the flight line. However, ambient white light requirements must be met by using black cloth or other light shields.

Extremely dark testing areas and high ultraviolet radiation levels are not always necessary. They are, however, always helpful and do ease the testing chore considerably. Further, in tests for the very smallest indications such as microcracks in jet turbine blades, they are necessary. In many instances, minute indications could only be detected by a dark adapted observer working in a booth with no more than 10 lx (1 ftc) of white light and then only by holding the part 5 or 7.5 cm (2 or 3 in.) from a 100 W mercury arc source, producing 150 to 180 W/m^2 intensities on the test object.

The window lens of the visible radiation (white light) sensor must not fluoresce. Under near ultraviolet radiation of 3500 $\mu W/cm^2$, visible radiation of 4.8 klx (440 ftc) has been reported instead of 13 lx (1.2 ftc) as measured with a photometer with a nonfluorescing sensor window.

The axiom that high ultraviolet radiation levels and low ambient light levels produce the best performance is still true. However, maximum performance is not always necessary, so tests can often be carried out under far from ideal conditions. Nevertheless, those ideal test conditions will considerably ease the test and should be used if at all possible.

As an example, an ultraviolet radiation illumination level of about 800 to 1000 µW/cm² is sufficient to reveal most anomaly indications. This will depend, of course, on the size of the indication and type of liquid penetrant system used. There will be applications where extremely high levels of ultraviolet radiation intensity will be required.

Light Meters

White light and ultraviolet radiation are measured with different sensors and in different units of measurement. A standard white light sensor will measure in footcandles or lux, and the ultraviolet radiation sensor will measure in microwatts per centimeter squared (µW/cm²). Both meter and digital type instruments are available, as shown in Figure 4.13.

Figure 4.13: Photoelectric instruments for measurement of ultraviolet radiation intensity: (a) handheld digital ultraviolet radiometer measures near ultraviolet radiation in 0 to 19 990 µW/cm² range; (b) radiometer/photometer with dual sensor for measuring ultraviolet radiation and visible (white) light; and (c) radiation monitoring and verification fixture ensures accurate measurement of ultraviolet radiation at 38 cm (15 in.).

(a) (b) (c)

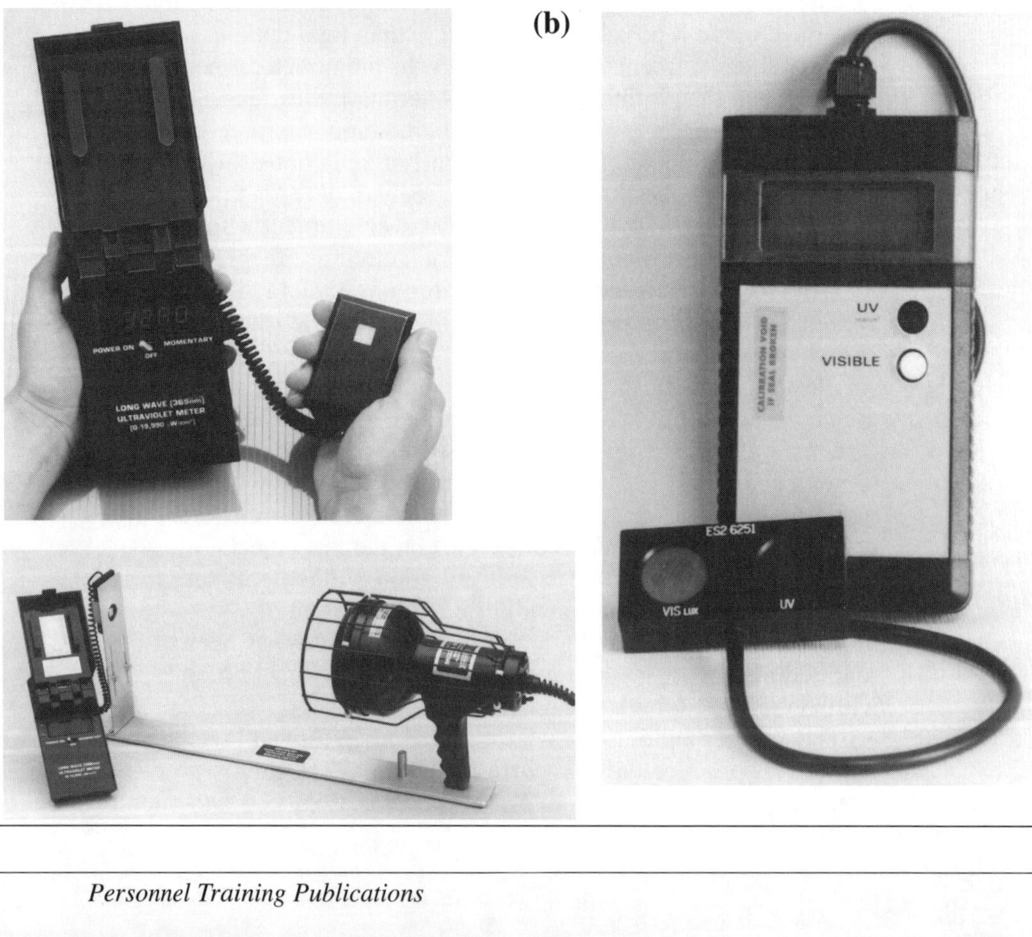

Materials for Liquid Penetrant Testing

The materials used in liquid penetrant testing include penetrants, emulsifiers, removers or cleaners and developers. Penetrants, emulsifiers and removers and/or cleaners are furnished in liquid form. Developers are furnished in powder form. The powders may be used in the dry state, or are mixed with a suitable liquid (usually water) before use. Most of the penetrants, cleaners and developers are available in pressurized spray cans, as well as in bulk. Concentrations, usage and maintenance are in accordance with the manufacturer's directions. All bulk materials in use require a daily check for contamination.

Precleaning and Postcleaning Materials

Special precleaning and postcleaning materials are not required for most test objects. However, the cleaning materials must be compatible with the penetrant. Manufacturers will recommend an approved cleaner for their products. Commercially available solvent cleaners are usually satisfactory. Some procedures require sulfur and halogen free cleaners for nickel and titanium alloys. Some test objects may have special cleaning requirements that will be specified in the company procedure.

Water Washable Penetrants

Water washable penetrants are highly penetrating, oily liquids containing a built in emulsifying agent that renders the oily media washable with water. The simplest to use are visible penetrants or color contrast penetrants, because no fluorescent lighting is required. They contain a dye (usually a bright red, but sometimes a special color such as blue) that can be seen under white (visible) light. Greater visibility is obtained if fluorescent penetrants are used and viewed under ultraviolet radiation. The color of fluorescence is usually a brilliant yellow-green. For special applications, there are fluorescent penetrants that glow red or blue.

Postemulsification Penetrants

Postemulsification and solvent removable penetrants have similar formulations to those of water washable penetrants, except they do not contain an emulsifying agent and consequently are not soluble in water. These penetrants must be treated with a separate emulsifier before they can be removed by a water rinse or wash, or they can be removed using an approved solvent remover or cleaner.

Postemulsification penetrants are available as either visible or fluorescent penetrants. These penetrants have the advantage of eliminating some of the danger of over rinsing.

Emulsifiers

When applied to a postemulsification penetrant, emulsifiers combine with the penetrant to make the resultant mixture water washable. The emulsifier, usually dyed orange to contrast with the penetrant, may be either lipophilic (oil base) or hydrophilic (detergent water base).

The lipophilic emulsifiers are usually used as contact emulsifiers. They begin emulsifying on contact with the penetrant. Emulsifiers can never be applied by brushing. After the proper emulsification time, the mixture of penetrant and emulsifier can be removed with a standard water rinse. Lipophilic emulsifiers remove excess surface penetrant, allowing the remaining penetrant to be water washable.

The hydrophilic emulsifiers (removers) can also be used as contact emulsifiers, but the emulsifier is diluted with water for tank use or sprayed as a water mix rinse. These penetrants have the advantage of eliminating some of the danger of over rinsing. Hydrophilic emulsifiers require agitation to allow fresh emulsifier to contact the surface penetrant.

Solvent Removers

Solvent removers or cleaners are used to remove excess penetrant from test surfaces. In selecting a solvent remover, only those materials approved by the penetrant manufacturer and penetrant procedure can be used. These solvents are available in aerosol cans or bulk.

Dry Developer

Dry developer is a fluffy powder that is applied to dry test surfaces (after the removal of excess penetrant) for the purpose of absorbing penetrant from discontinuities and enhancing the resultant indications. Of the different developers available, dry developer is the most adaptable to rough surfaces and automatic processing. It is also the easiest to remove. Dry developer sensitivity is less than the other types of developers, but it is better suited for certain applications, such as fine threads or rough castings, and provides better resolution. Dry developer is also much easier to remove after the test.

Nonaqueous Wet Developer

Nonaqueous wet developer is a suspension of developer particles in a rapid drying solvent. It is most often used with solvent removed processing. Like dry developer, nonaqueous wet developer is applied only to dry surfaces. Of all the developers, the nonaqueous wet developer is the most sensitive in detecting fine discontinuities. The evaporation of the solvent carrier helps to draw the penetrant from discontinuities.

Water Based Developers

Water based developers function similarly to dry developer, except they are applied before drying the test object. Two types of developer are available. In the water suspendible developer, particles are held in suspension in water and require continuous agitation to keep the particles in suspension. In the water soluble developer, powder is dissolved in water, forming a solution; once mixed they remain mixed. Water based developer requires a periodic check (usually weekly) for concentration with a hydrometer.

Special Purpose Penetrant Materials

In addition to the conventional penetrants, emulsifiers, removers and developers used in liquid penetrant testing, there are low sulfur and low chlorine materials for testing nickel alloys, certain stainless steels and titanium. Special purpose inert materials are available for testing objects that come in contact with liquid oxygen, rubber or plastic.

Food compatible materials are also available. There are high temperature penetrants for testing hot welds, and special penetrants for testing at low temperatures. There are supersensitive penetrants for detecting extremely fine discontinuities, and penetrants that provide sufficient contrast and sensitivity without a developer.

There are low energy emulsifiers and inhibited solvent removers to slow emulsification and the removal of excess penetrant. There are also wax and plastic film developers that absorb and fix penetrant indications to provide record of the test.

The selection and usage of these materials is largely dependent on the particular process used and the controlling specifications or standards. All special purpose materials require specific approval for industrial applications.

PRECAUTIONS

The liquid penetrant testing technician must be sure that the test object is not damaged or overheated during the test. If any test object contains plastic, rubber or other nonmetallic components, the technician must check to see if solvent will harm that test object. Liquid penetrant materials and solvents require standard safety procedures, as detailed by their manufacturers. There are also ecological requirements for disposal of penetrant residue and rinse water. The solvents and aerosol cans are flammable, and normal precautions for flammable liquids apply.

PTP

The ASNT PERSONNEL TRAINING PUBLICATIONS

PT

Level II

Chapter 5

Selection of Liquid Penetrant Testing Method

Selection

Selection of penetrant material combinations of penetrant, emulsifiers, penetrant removal methods and forms of developers must be made in accordance with the penetrant material manufacturer's recommendations and instructions or an applicable qualified products list. The required liquid penetrant type and method are sometimes decided by the contractor, customer, Level III technician or design engineer. In some cases, managers have the option of using different combinations of penetrants, emulsifier systems, penetrant removal methods or developers.

Selection of a suitably efficient, cost effective penetrant type and process for a particular application depends on several factors.

1. Customer requirements.
2. Specification requirements.
3. Sensitivity required.
4. Size and number of test objects.
5. Surface condition of the test objects.
6. Configuration of the test objects.
7. Cost of equipment and materials.
8. Availability of water, electricity, compressed air and suitable testing area.

Advantages and Disadvantages

The majority of liquid penetrant tests can be done effectively and efficiently using Type I (fluorescent), Method A (water washable) with a sensitivity Level 2 or 3. This is true for large castings, forgings or small, machined test objects. For aircraft engine parts, a Level 3 or 4 hydrophilic system is recommended. To operate efficiently, the penetrant system must be the proper size with the proper handling equipment.

The advantage of the fluorescent penetrant is that fluorescent indications are easier to see. The advantage of the water removable penetrant is that it is much faster, safer, ecologically friendly and inexpensive. The disadvantage of water removable penetrant is the danger of over washing if the technician does not use proper techniques. Over cleaning can occur with any type, method or sensitivity level if technicians fail to follow the appropriate work practices. The advantages and disadvantages of the different types and methods are listed in Table 1.1.

Table 1.1: Advantages and disadvantages of the different types and methods of liquid penetrant testing.

Technique	Advantages	Disadvantages
Type I	Easier to see indications	Dark area and special lighting required
Type II	No special lighting required	Difficult to see indications
Method A	Fast	Danger of over washing
Method B and D	Can achieve sensitivity Level 4	Extra step and material required
	Less danger of over washing	Slower
Method C	Portable and no power required	Slower

Portability

The majority of onsite field liquid penetrant tests can be effectively and efficiently completed using portable, Type I (fluorescent) or Type II (visible) penetrant kits and either water or solvent wipe. The aerosol liquid penetrant testing kits are portable, and if Type II solvent removable is chosen, no power or water supply is needed.

The advantage of aerosol kits is their portability, and very little set up time is required. The disadvantage of manual wipe is that the method is slow.

Postemulsification

When postemulsification processes are required, appropriately formulated penetrant materials and postemulsification methods must be used. Small test objects can be dipped in a tank of Method B (lipophilic) emulsifier, the emulsification time monitored and excess penetrant rinsed off. For large test objects, equipment that sprays water and Method D (hydrophilic) emulsifier mix may be more efficient.

The advantage of the postemulsified method is the elimination of worry that penetrant may be rinsed out of shallow discontinuities.

The disadvantage of the postemulsified method is the cost of the emulsifier and equipment, the extra timed step for emulsification and the maintenance checks for the emulsifier. Also, the Method B (emulsification) method is not permitted by some industry specifications for aircraft engine parts. Choice of developer for all the methods is regulated by specifications and the manufacturer's recommendation.

Dry Developer

When a dry developer is used, the test objects must be completely dry before application. Dry developer is best for rough surfaces, such as castings or test objects with fine threads or corners such as keyways. Dry developer is also easy to handle and easy to apply.

Dry developer leaves no film, thus no special cleaning is required for subsequent processing operations. Excessive drying time should be avoided, as this will reduce sensitivity. Dry developer is the least sensitive type of developer and should not be used for visible penetrant.

Wet Developers

Aqueous developers can be applied immediately after the water rinse, before drying. Both water suspendible and water soluble developers may be applied by dipping, spraying or flowing. Aqueous developers give off no flammable vapors, no dust and have no odor. Since aqueous developer leaves a thicker coverage, complete coverage with aqueous developer is easier to monitor than with dry developer. Both types of wet developers leave a thicker buildup of developer on rough surfaces, such as castings or test objects with fine threads or corners such as keyways. The water suspendible developer requires agitation, and both types require daily maintenance checks when used in tanks.

When a nonaqueous wet developer is used, the test object must be completely dry before the developer is applied. Nonaqueous wet developer is the most sensitive developer because it can be sprayed in a very controlled, thin layer. The solvent carrier dissolves in the penetrant, slightly reducing the viscosity, increasing the volume of the penetrant in the discontinuities and leaching the indications to the surface as the volatile liquid evaporates.

There are also wax and plastic film developers that absorb and fix penetrant indications to provide records. The selection and usage of these materials is largely dependent on the particular process used, the controlling specifications or standards and the company policy regarding recording of test indications.

Chapter 6

Interpretation and Evaluation of Indications

Discontinuity Categories

Specific discontinuities are divided into three general categories: inherent, processing and service. These categories are further classified as to stage of manufacturing, the material (ferrous or nonferrous) and the manufacturing process. No matter what type of discontinuity may be present in the test object, only those open to the surface may be detected with liquid penetrant testing.

Inherent Discontinuities

In metals, inherent discontinuities are those that are related to the melting and original solidification of the molten metal, ingot or casting.

1. Ingot discontinuities are those related to the melting, pouring and solidification of the original ingot, and include hot top shrinkage, slag, porosity, cracks and nonmetallic inclusions.
2. Inherent cast discontinuities are those related to the melting, casting and solidification of the cast test object. They include those discontinuities introduced by casting variables, such as inadequate feeding, gating, excessively high pouring temperature, entrapped gases, shrinkage, hot tears, inclusions, cracks and blowholes.

Processing Discontinuities

Processing discontinuities are related to the various manufacturing processes, such as forging, machining, forming, extruding, rolling, welding, heat treating and plating. These discontinuities may be caused by the processing method, or they may be inherent discontinuities that have been exposed or changed (in shape or direction) by the processing method.

Service Discontinuities

Service discontinuities are related to service conditions such as cycles of loading, stress corrosion, fatigue and wear.

Forming of Discontinuity Indications

Discontinuities that have an opening to the surface can be detected using the liquid penetrant method. After the proper precleaning, drying and application of penetrant for the proper dwell time (typically about 10 min), capillary action forces the penetrant into the discontinuity. Then the excess surface penetrant is removed either by wiping or with a water spray rinse, with the aid of an emulsifier for postemulsifiable penetrants.

Time for Indications to Appear

After removal of excess surface penetrant and application of developer, penetrant will migrate to the surface aided by capillary action and the blotting action of the developer. Deep indications will begin to appear first since indications with a large reservoir of penetrant will bleed out faster than a small indication. The typical time to allow indications to form before measuring and evaluation is specified in the written procedure. Standard times for evaluation are important so different technicians and testing facilities will get standard results.

Persistence of Indications

One method to verify that a questionable indication is a relevant indication is to wipe the penetrant away with a swab dampened with solvent. The persistence of a relevant indication will cause it to bleedout again. Caution must be used to ensure that the swab is only dampened and not dripping with solvent.

Effects of Temperature

The standard temperature range for penetrant testing is 10 to 38 °C (50 to 100 °F). Colder temperatures will increase the viscosity of the penetrant and slow the capillary action. Some specifications allow temperatures of 4.4 to 65.5 °C (40 to 150 °F), but require doubling the dwell time for temperatures below 10 °C (50 °F).

Temperatures higher than approved may cause the penetrant to dry and reduce sensitivity. Some specifications allow special, high temperature materials that are available for temperatures up to 177 °C (350 °F). High temperature materials require special procedures, special training and certification of technicians.

Lighting

The standard lighting for viewing and evaluating visible dye penetrant indications is a minimum of 1000 lx (100 ftc) of white light at the test surface. For fluorescent penetrant indications, the standard lighting is a minimum of 1000 μW/cm^2 ultraviolet radiation at the test surface, and a darkened test area of less than 20 lx (2 ftc).

The standard lighting in a fluorescent rinse or penetrant removal area is less than 100 lx (10 ftc) and greater than 100 μW/cm^2.

Effects of Metal Smearing Operations

Operations such as power wire brushing or sand blasting can smear metal and close the surface opening of discontinuities. If these operations have been performed on a test object, the surface must be etched to remove the smeared metal. Samples of the visual appearance of smeared metal should be available or used in technician training.

Sequence

Inservice tests are sometimes used for the root layer of welds, or before final machining. Final liquid penetrant testing is performed on a test object in the final machined and heat treated condition after proper precleaning.

Test Object Preparation

The test object must be processed through preparation, such as etching, precleaning and drying, as efficiently as possible.

Factors Affecting Indications

Penetrant Used

The choice of penetrant depends on the sensitivity requirements, location, availability of ultraviolet radiation, water, power and the requirements of the inspection. Fluorescent indications are easier to see, and the postemulsified method is the most sensitive for small shallow indications.

Prior Processing

The method of precleaning may depend on prior processing of the test object. If a test object or weld has never been exposed to machine oils or lubricants, some specifications will allow reduction of penetrant dwell time to as little as 5 min. Ultrasonic testing using couplant should only be performed after final penetrant testing because the couplant may hinder penetrant penetration into discontinuities. Visible penetrant should never be used before a fluorescent penetrant test because the red penetrant may still be present in discontinuities and could interfere with indication luminosity.

Crack Indications

Cracks can occur at any stage of manufacture or service. An indication from a crack will be linear (three times as long as wide) and slightly irregular. Test objects that have cracks are usually rejected.

Solidification Cracks

Ingot cracks occur during metal solidification, and may change to a straight line during processing. They appear as straight line indications in the direction of rolling or working.

Processing Cracks

A typical processing crack can occur during welding, straightening, bending or from internal stresses at any point during processing, including heat treatment or plating. These cracks will also be linear and slightly irregular, and test objects that have processing cracks are routinely rejected.

Service Cracks

Cracks can occur at any time during service for a multitude of reasons. Usually they are caused by overloading, fatigue or, in some cases, corrosion. Test objects that have inservice cracks are typically rejected.

Porosity Indications

Liquid penetrant indications from porosity are rounded (less than three times as long as wide) and can occur in the ingot, casting or weld. Typically no porosity is present or allowed in forged, rolled or extruded materials. Some small porosity is acceptable to some specifications, especially for castings and welds.

Indications from Specific Material Forms

Forgings

Laps are caused when metal is folded over and not fused either at transition areas or from improper trimming at flash line areas. Laps will be open to the surface, linear (three times as long as wide) with some shape but no particular orientation to the direction of forging.

Bursts are caused by forging at an improper temperature. Bursts can be at the surface if the surface temperature is incorrect, or internal if the internal temperature is incorrect. Bursts will be linear and branch or chevron shaped.

Most critical forgings are tested for internal discontinuities with ultrasonic testing. Liquid penetrant is typically used for nonmagnetic materials. Magnetic particle tests are often used for ferromagnetic forgings to detect surface indications.

Castings

Porosity or trapped gas pockets that couldn't escape at the risers can form either at the surface of within the interior of the test object. Surface connected porosity will produce rounded indications.

Shrinkage is caused by stresses from unequal cooling. Surface shrinkage can be detected visually or by liquid penetrant testing of the surface. Shrinkage is normally internal and detected by radiography, unless shrinkage is machined open to the surface.

Hot tears are caused by unequal cooling between thick and thinner sections of castings. Indications will be linear and crack like, but usually wider than a crack. Hot tears are usually open to the surface, but may be inaccessible for liquid penetrant testing. Radiographic testing can detect hot tears.

Cold shuts are caused by metal solidification before the metal melts together. This is caused by splashes or streams of molten metal meeting and solidifying, instead of flowing together. Cold shuts may be open to the surface. They can be detected by radiographic testing.

Critical castings are often tested with radiography for internal discontinuities because the large grain sizes make ultrasound difficult. Liquid penetrant is used to detect surface discontinuities in nonmagnetic castings, and magnetic particle testing is used for ferromagnetic castings.

Plate

Laminations are the most common form of discontinuities in plate. They are caused by relatively large discontinuities, such as blow holes, slag or inclusions, which are rolled flat and parallel to the surface and are open to the surface at the end or edge of the plate. Ultrasonic testing is used to detect discontinuities in critical plate. Some laminations are acceptable if they are outside a weld area and meet certain spacing and size criteria.

Welds

Porosity is a discontinuity in welds that can be open to the surface or subsurface. Surface porosity will appear as a round indication. Porosity can be caused by improper welding technique or moisture, oil or foreign material in or near the weld metal.

Crater cracks or crater pits are in the weld crater. They are found where the welder stopped a length of weld and used an improper technique for stopping. Crater pits may go to the root of the weld, and both conditions will result in a reject liquid penetrant indication.

Weld fusion lines can be difficult to clean properly if acceptable weld undercut is present. Weld undercut occurs where the weld has not completely refilled the edge of the melted base material. This area is where fusion line cracks can also occur, so care must be used when evaluating this area. In some cases, additional grinding must be done on visually acceptable undercut to allow proper penetrant testing.

Transverse weld cracks can also occur usually from a filler material problem or hydrogen embrittlement.

Extrusions

Discontinuities in extrusions and in rolled shapes are usually longitudinally aligned, parallel to the direction of working. Acceptance criteria for forgings are used.

Evaluation of Indications

To evaluate an indication means to decide if the indication is acceptable, requires rework or causes the test object to be rejected. Most industry acceptance standards are written in specifications, and the design engineers predetermine what discontinuities are acceptable, and which ones require the test object to be scrapped or reworked. For some critical test objects, a technician may be told that no indications interpreted as cracks are allowed.

The evaluation first determines if the indication is linear (three times as long as wide) or rounded. Then the technician determines if the indication is a crack, seam, lap, porosity or lack of fusion, depending on knowledge of the test object's manufacture or history.

The next step may be to measure each indication and fill out a detailed report, including a sketch to submit for further engineering evaluation. It is important for a company to establish a system to ensure that rejected test objects are marked, separated from acceptable test objects and destroyed so they can never be used accidentally.

False Indications

The most common source of false indications is poor washing of water washable and postemulsified penetrants. When using fluorescent penetrant, ultraviolet radiation during the washing process is important. The technician must determine whether a good rinse is obtained or whether patches of fluorescence remain on the test object. Adequate lighting at a fluorescent rinse or cleaning station is usually considered 100 lx (10 ftc) visible light and 100 $\mu W/cm^2$ ultraviolet radiation at the test surface.

Solvent removal is slower and more thorough. To guard against confusion resulting from fluorescent or color spots other than relevant indications, care is taken so that no outside contamination occurs. Typical sources of contamination include the following.

1. Penetrant on the hands of the technician.
2. Contamination of wet or dry developer.
3. Penetrant rubbing off of one test object to a clean portion of the surface of another test object.
4. Penetrant spots on the testing table.

To avoid contamination, process tanks and testing areas are kept clean, only lint free wiping cloths are used and test objects are kept free of fingerprints and tool marks.

Relevant Indications

Relevant indications are those caused by a material discontinuity. The interpretation of an indication as relevant is a matter of observing the indication, eliminating the possibility of it being a false indication and then further determining that it is not a nonrelevant indication. Any relevant indication immediately becomes subject to interpretation of its cause (type of discontinuity).

Nonrelevant Indications

Nonrelevant indications are caused by features in the test object that are there by design, but are in no way a relevant discontinuity. Most nonrelevant indications are easy to recognize because they are related directly to some feature of the assembly that accounts for their presence.

Nonrelevant indications include those that appear on test objects that are press fitted, riveted or spot welded together, and those caused by surface roughness.

Any nonrelevant indication that interferes or could mask a relevant indication must be further evaluated and sometimes retested. If a test object is too rough to perform a proper liquid penetrant test, or has a condition like weld undercut that has been accepted visually, the test object or weld may have to be returned for further preparation for liquid penetrant testing.

Chapter 7

Liquid Penetrant Process Control

Quality Control of Test Materials

The reliability of any penetrant test is determined in large part by the condition of the materials used. Even the best procedures are ineffective if test materials are faulty. To ensure the satisfactory condition of penetrant test materials, various quality control procedures are used.

This chapter discusses the inservice checks used to test penetrant materials held in open tanks and subject to contamination or evaporation. Each of the quality control procedures mentioned here is based on the assumption that care, handling and use of the materials are strictly in accordance with manufacturers' recommendations.

Many additional quality control procedures, such as the determination of sulfur and chlorine content, liquid oxygen compatibility, temperature stability, water tolerance, viscosity, flash point, toxicity and developer precipitation rates, are described in the various controlling specifications and standards. These procedures, however, are primarily of interest to the manufacturer or laboratory technician rather than the individual performing or monitoring penetrant tests. Quality control procedures are also available from manufacturers and various technical societies such as the American Society for Testing and Materials (ASTM).

Test Material Control Samples

Many of the tests used in the control of penetrant materials are comparison tests in which used materials are compared with new materials. For this reason, control samples are taken at the time the materials are received from the supplier. These samples are kept in sealed containers and stored where they are not subject to deterioration from heat, light or evaporation. The procedures for taking the samples and their renewal are typically detailed in company procedures, controlling specifications and standards.

Reference Blocks

The use of reference blocks, plates or panels is often specified in the procedures for quality control of liquid penetrant materials. These standards are often called *known defect test specimens* (KDTS). The materials used in the manufacture of reference blocks include aluminum, steel, nickel, glass and ceramic. Some of the blocks are designed primarily for checking penetrant or system sensitivity and performing comparison tests, whereas others are designed specifically for testing penetrant or emulsifier washability. All, however, are prepared to rigid specifications as required for specific tests, as described below.

System Monitor Panels

Liquid penetrant system monitor panels, as shown in Figure 7.1, are commercially manufactured and can be used to detect major changes both in visible and fluorescent systems, and in water washable and postemulsified systems. The liquid penetrant system monitor panel is normally processed at the beginning of each shift to verify system performance.

The panel is stainless steel, about 0.25 cm (0.1 in.) thick and 10 by 15 cm (4 by 6 in.). A strip of chrome plating runs the length of the plate, and the other half of the same side of the plate is an oxide grit blasted surface to monitor background color or fluorescence. The chrome strip has five variable size crack centers induced by a hardness punch from the reverse side. The largest crack pattern is readily visible with low sensitivity penetrant materials, and the smallest is difficult to observe even with high sensitivity penetrant materials.

Figure 7.1: A liquid penetrant system monitor panel includes five crack centers of different sizes for evaluating sensitivity and includes a grit blasted section for judging wash characteristics of the liquid penetrant system: (a) drawing; (b) under white light; and (c) under ultraviolet radiation.

(a) **(b)** **(c)**

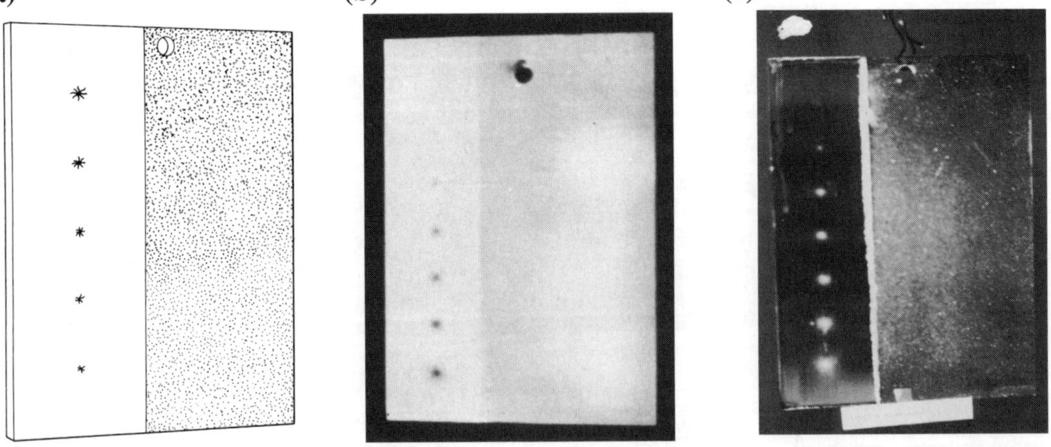

Aluminum Reference Blocks

Aluminum reference blocks measure about 5 by 7.5 cm (2 by 3 in.), and are cut from 0.8 cm (0.3 in.) thick bare 2024-T3 aluminum alloy plate, with the 7.5 cm (3 in.) dimension in the direction of rolling. The dimensions are for guidance only and are not critical.

The blocks are heated and water quenched to produce thermal cracks. This is accomplished by supporting the block in a frame and heating it nonuniformly with the flame of a gas burner or torch in the center on the underside of the block. The flame remains centered and does not move in any direction during the heating process.

A 510 to 527 °C (950 to 980 °F) temperature indicating medium – typically in the form of lacquer, pellets or sticks – is applied to an area the size of a penny on the top side and directly in the center of the block. The heat of the torch or burner is adjusted so that the block is heated about 4 min before the temperature indicating medium melts, after which the block is immediately quenched in cold water, as shown in Figure 7.2.

A groove about 0.15 by 0.15 (0.06 by 0.06 in.) deep is cut in the 5 cm (2 in.) direction across the center of the heat affected zone. This forms two test areas, and permits the side-by-side application and comparison of two penetrants without cross contamination. This type of block is widely used for comparing the performance of penetrants in field conditions.

Figure 7.2: Heating and quenching of an aluminum reference block.

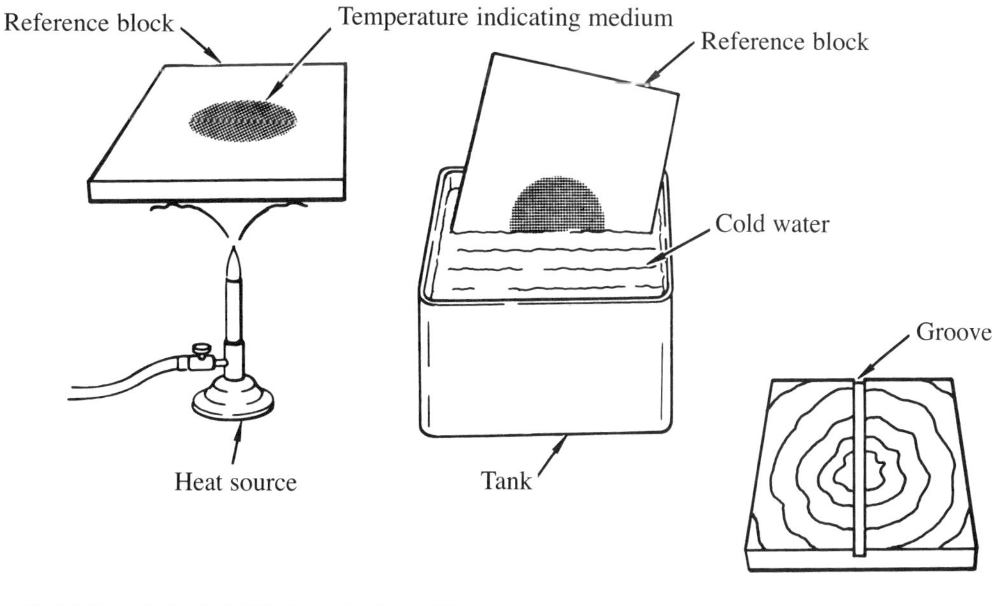

Preparation for Use

Before use, aluminum reference blocks are scrubbed with a bristle brush and liquid solvent, followed by degreasing and drying.

Preparation for Reuse

After a reference block has been used, it is cleaned before reuse. The block is heated slowly with a gas burner to 426 °C (800 °F), as determined by a 426 °C (800 °F) temperature indicating medium, after which the block is quenched in cold water. It is then heated to about 71 °C (160 °F) for 30 min to drive off any moisture in the cracks, and is then allowed to cool to room temperature.

Ceramic Reference Blocks

Ceramic reference blocks are flat, circular disks of unglazed ceramic that, although quite solid and impervious to liquids, have micropit surfaces that entrap liquid penetrants. The micropit structure provides a range of pore sizes, and a performance comparison can be made of two or more penetrants merely by noting the number or distribution of porosity indications and their intensity in a side by side comparison test. Indications appear as a large number of microscopic specks of fluorescence or color, the number increasing as the sensitivity of the penetrant increases.

Preparation for Use

The ceramic reference block is removed from its container (a shallow jar containing alcohol) using a pair of tweezers to avoid touching with the fingers. The block is then wiped with a soft absorbent tissue, and allowed to dry for 5 min.

Usage

Using a small applicator, a drop of each penetrant to be tested is applied to the flat surface of the reference block. Immediately following application, the penetrants are blotted with a piece of soft tissue by pressing the tissue against the block using a flat object. The tissue and cover also hold the penetrants in proper contact with the reference block and prevent excessive bleeding and possible cross contamination.

Following the required dwell period (usually 10 min), the penetrants are processed in accordance with the penetrant manufacturer's recommendations. A developer, however, is not used. In making visual comparisons, both the number of indications observed and the intensity of indications are noted.

Cleaning After Use

After use, the ceramic block is placed on edge in alcohol and soaked for several hours or until the penetrant entrapments diffuse into the alcohol. The block is then returned to its storage container, covered with fresh alcohol and sealed.

Anodized and Plated Test Panels

Stress cracked anodized aluminum and chrome plated nickel test panels are frequently used for comparing penetrant sensitivity and washability. The panels are classified according to the size cracks they contain. The grades are coarse, medium and fine, providing low, medium and high sensitivity levels (see Figure 7.3).

Figure 7.3: Fluorescent indications of stress cracks in chrome nickel test panels: (a) coarse cracks; (b) medium cracks; and (c) fine cracks.

(a) **(b)** **(c)**

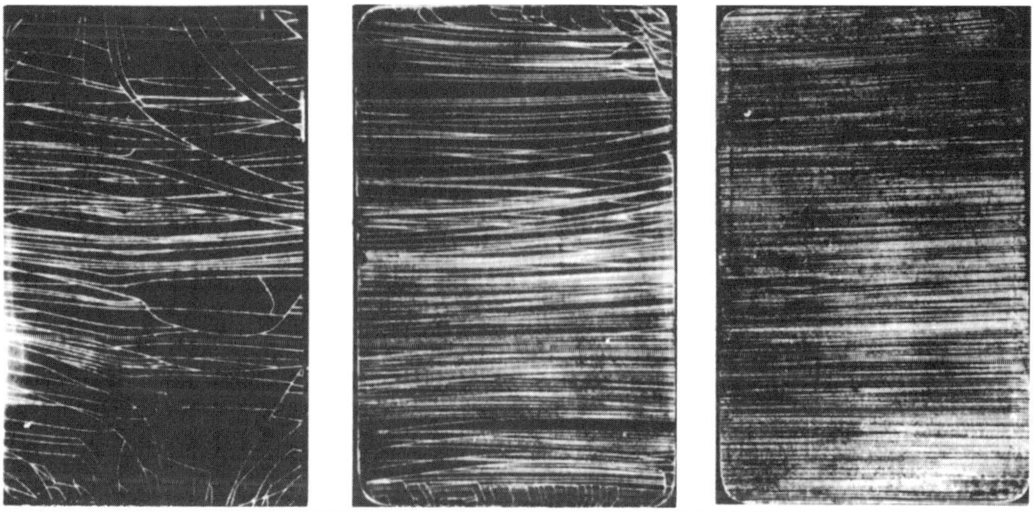

Preparation for Use

Panel surfaces are scrubbed using moderate pressure and a cloth dampened with emulsifier or concentrated soap solution, followed with a thorough rinse using a water spray. After rinsing, the panel is oven dried at an approved temperature for 3 to 5 min to drive off moisture remaining in cracks. The panel is then dipped in acetone and agitated for about 1 min, removed and again oven dried at an approved temperature for about 5 min.

Usage

A line is usually drawn along the centerline (length) of the panel using a wax pencil or narrow vinyl tape. This forms two test areas and permits the side by side application and comparison of penetrant materials without cross contamination.

Twin NiCr Sensitivity Panels

A set of two NiCr panels, each measuring 10 by 3.5 cm (3.9 by 1.4 in.) is sheared from the same stock. This allows for matching crack patterns.With a matched set of panels, a simultaneous comparison of two penetrant materials can be achieved. The panels come in sets having crack depths of 1 cm (0.4 in.), 2 cm (0.8 in.), 3 cm (1.2 in.) and 5 cm (2 in.).

Stainless Steel Test Panels

Stainless steel test panels or plates are used in testing the washability of fluorescent penetrants and visible dye penetrants. The panels are prepared from annealed type 301 or 302 stainless steel and measure 5 by 10 cm (2 by 4 in.) or larger. Each is sandblasted on one side with 80 mesh average size grit, using 60 lb air pressure (414 kPa), with the gun held about 46 cm (18 in.) from the surface. The sandblasting is continued until a uniform matte surface is obtained. It is the sandblasted surface that is used in performing tests.

Preparation for Use

Before use, the panels are cleaned by degreasing, heated to 71 °C (160 °F) and then allowed to cool to room temperature in a dry area.

Low Cycle Fatigue Blocks

Titanium or NiCrFe plates are commonly used to manufacture standards with low cycle fatigue blocks (LCF) cracks in various size ranges. The cracks are started from electrical discharge machined (EDM) notches or spot welds, which are later ground away after the starter cracks are grown. Tensile stressing or reverse bending of the plates achieves additional crack length extension. Titanium or NiCrFe plates are commonly sold in a set of three plates, with a total of eighteen possible cracks in the set.

Low cycle fatigue blocks are used like other known discontinuity standards, except that the total number of detected cracks per inspection of the plate set is recorded and monitored in a running summary per procedure supplied with the plates. When fewer cracks are detected, the technician is warned that something has shifted in the process capability, or that the cracks have been improperly cleaned.

The low cycle fatigue blocks are cleaned and stored in acetone or equivalent solvent or in a foam form made for them, as long as they are cleaned with alcohol or acetone before and after each use. The low cycle fatigue blocks can also be returned to the manufacturer for extensive cleaning, if necessary.

Tests of Liquid Penetrant Materials

The inservice quality of the materials used in liquid penetrant testing is determined by a check of sensitivity, water content, contamination and washability. The tendency toward fading of the penetrant dyes is also checked by a simple comparison test. The tests described below are typical of those performed on used or questionable penetrants. The tests should be performed by laboratory technicians.

Sensitivity Comparison Test

When performing a sensitivity comparison test, the penetrant is applied to one half of the reference block, and the reference or control penetrant is applied to the remaining half. The processing used, including dwell time, emulsification or removal and developing, is recommended by the penetrant manufacturer.

The indications are then visually compared under the appropriate lighting (normal or white light for visible dye penetrant indications and ultraviolet radiation for fluorescent indications). If a noticeable difference exists in the sensitivity or intensity of indications (as determined by visual observation), the penetrant is discarded. Likewise, if the penetrant shows evidence of contamination from dirt, it is discarded.

For the visual examination of fluorescent indications, the ultraviolet radiation source must have an intensity of at least 1000 $\mu W/cm^2$ at the test surface.

Water Washability Test

To achieve maximum contrast between indications and background, excess surface penetrant must be readily removable. In the water washability test, the performance of the penetrant is compared to that of the reference penetrant.

The penetrants are applied to separate test panels. After normal dwell and draining periods and emulsification (if applicable), the penetrants are washed from the panels using a uniform water spray. If washing is found to be difficult, or retention of background penetrant is noticeably different from that of the reference penetrant, the penetrant is discarded.

Water Content Test

Water content of a penetrant is best determined by the test described in ASTM *Standard D95-99e1: Standard Test Method for Water in Petroleum Products and Bituminous Materials by Distillation*. About 100 mL of the penetrant is placed in a boiling flask with a similar quantity of moisture free xylene. The flask is connected to a reflux condenser so that the condensate drops into a 25 mL graduated tube where the water settles out. When no more water is being gathered in the graduated tube (usually after a period of 1 h), the boiling process is terminated. After cooling, the volume of water in the graduated tube is read. The volume in milliliters is the percent of water by volume present in the penetrant.

The penetrant is also examined for evidence of gelling, separating, clouding, coagulating or floating of water on the penetrant surface. If any of these conditions exist, or the percent of water exceeds specification requirements, the penetrant is discarded.

Fluorescent Luminance Test

The fluorescent luminance test is a comparison of the luminance of the control penetrant sample to the used penetrant in the tanks. If a visual comparison is made, differences of 25% are obvious, differences of 12% are noticeable and differences of 6% are detectable by the eye. Experts may sometimes detect 3% differences, but these are not typically detectable to the average observer. Most specifications require the luminance test be performed with a fluorometer.

In this test, a small amount of the penetrant to be tested and the reference penetrant are diluted with a nonfluorescent, highly volatile solvent such as acetone. Test papers, cut to fit the sample holder of the fluorometer, are then dipped into the solutions, withdrawn and dried. The samples are then alternately read on the fluorometer and the results compared. If the fluorescent intensity of the penetrant should drop below 90% of the reference penetrant, the penetrant is discarded.

Emulsifier Tests

Emulsifiers are usually tested for their sensitivity, washability, water content and the amount of contamination from penetrants. As a general rule, the tests should be performed by laboratory technicians in accordance with ASTM specifications.

The sensitivity test, the water washability test and the water content test for emulsifiers are identical to the tests used for penetrants.

Developer Tests

Dry Developer

Dry developers used in open containers or chambers are usually tested by observation. Because they are not hygroscopic, they do not absorb moisture from the air and they are relatively trouble free if they do not come in contact with water. Any dry developer that is found lumpy or caked instead of light and fluffy, or that shows any other sign of having been wet, is discarded.

Dry developer is checked daily for cracking. Additionally, the developer is visually examined for dirt. It is also checked under ultraviolet radiation for fluorescent penetrant contamination or dirt by spreading a thin layer of penetrant in a 10 cm (4 in.) diameter circle. Should either condition exist, the developer is discarded.

Wet Developer

Wet (aqueous) developers are usually tested for proper concentration and possible contamination from dirt or penetrant. Solution concentration is tested by measuring the specific gravity with a hydrometer, as shown in Figure 7.4. If the hydrometer reading differs from manufacturers' requirements, either powder or water is added to the developer in sufficient quantities to bring the concentration within acceptable limits. Additionally, a small smooth test plate with no indications is dipped in the developer tank and visually examined for dirt.

The tank and the plate are checked for fluorescence or penetrant contamination under ultraviolet radiation. If either condition is in evidence, the developer is discarded.

Figure 7.4: Typical hydrometer (shown horizontally, but used vertically).

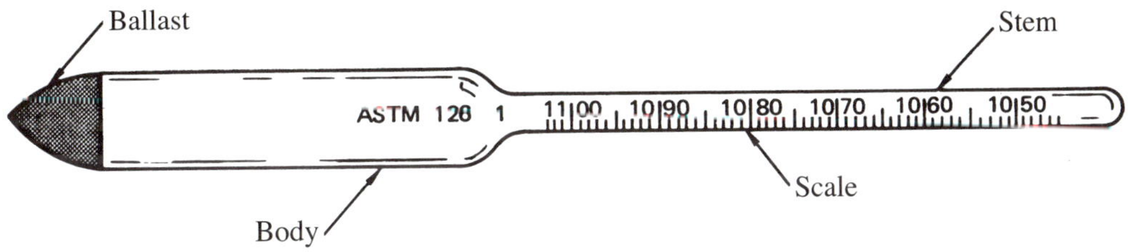

Chapter 8

Test Procedures and Standards

PROCEDURES, STANDARDS AND CODES

There are several different standards for liquid penetrant testing in industrial construction, manufacturing and maintenance applications. There are minor differences in these standards that are easily incorporated into a single procedure that covers the two types and four methods of liquid penetrant testing.

Acceptance Criteria

Typical liquid penetrant acceptance criteria are as follows (sizes will vary depending on test object application).

1. No linear indications allowed.
   1.1. A linear indication is three times as long as it is wide. Some specifications specify that all indications less than 0.15 cm (0.06 in.) are evaluated as rounded indications. Some specifications specify that all indications 0.05 cm (0.02 in.) and smaller are disregarded.
2. No rounded indications greater than 0.08 cm (0.03 in.) for materials up to 0.6 cm (0.25 in.) design material thickness, and no rounded indications greater than 0.23 cm (0.09 in.) for materials greater than 0.6 to 1.3 cm (0.25 to 0.5 in.) design material thickness. No rounded indications greater than 0.1 cm (0.3 in.) for materials with a design material thickness greater than 1.3 cm (0.5 in.).
3. The total area of acceptable rounded indications shall not exceed 0.375% of the total area of the test object or weld in any 255 cm^2 (6 $in.^2$) area, and there shall be no more than five indications in any 6.25 cm^2 (1 $in.^2$) area.
4. Indications separated by less than the length or diameter of the largest indication shall be considered a single indication measured from the extremities of the two indications.

Basic Methods of Instruction

Each facility should evaluate the methods of instruction and classes that are required for their liquid penetrant testing trainees, Level I, Level II and Level III technicians and other personnel involved with the penetrant process. Training methods may include formal classroom lectures, video presentations, informal demonstrations and hands on processing of test objects.

Experience has shown that repetition of key points, hearing and reading the same concepts and hands on practice are valuable training methods. All training for eventual certification should be documented in the company written practice as described in The American Society for Nondestructive Testing's *Recommended Practice No. SNT-TC-1A*, with a formal course outline and the hours documented for each trainee. A method for documentation of work experience hours must also be completed before certification to Level I, Level II or Level III.

Conclusion

Nondestructive testing students completing this classroom training book will have taken an important step, from unaided visual testing to that of a nondestructive testing method that is capable of illuminating discontinuities invisible to the unaided eye.

Additionally, this classroom training book provides background on critical practices for control and qualification of applied liquid penetrant technology.

Students may add technical knowledge in other nondestructive testing technologies to this basic understanding of process qualifications and controls. The results are the uniform and repeatable tests that are a hallmark of an individual certified as a Level II in liquid penetrant testing.

Glossary

Aqueous developer: See **Wet developer**.

Background: The surface on which an indication is viewed. It may be the natural surface of the test object, or it may be the developer coating on the surface. This background may contain traces of unremoved penetrant (fluorescent or visible), which if present, can interfere with the visibility of indications.

Background fluorescence: Fluorescent residue observed over the general surface of the test object during a fluorescent penetrant test.

Bath: Term used colloquially to designate the liquid penetrant testing materials into which test objects are immersed during the testing process.

Black light: See **Ultraviolet radiation**.

Bleedout: The action of the entrapped penetrant in spreading out from surface discontinuities to form an indication.

Blotting: The action of the developer in soaking up penetrant from a surface discontinuity, so as to cause maximum bleedout of penetrant for increased contrast and sensitivity.

Capillary action: The tendency of liquids to penetrate or migrate into small openings, such as cracks, pits or fissures.

Clean: Free of interfering solid or liquid contamination.

Color contrast penetrant: See **Visible dye penetrant**.

Comparative reference block: An intentionally cracked metal block having two separate but adjacent areas for the application of different penetrants so that a direct comparison of their relative effectiveness can be obtained. Can also be used to evaluate penetrant test techniques and test conditions.

Contact emulsifier: A fluid that begins emulsifying penetrant on simple contact with the penetrant. Usually oil based (lipophilic).

Contrast: The difference in visibility (lumination or coloration) between an indication and the surrounding surface.

Dark adaptation: The adjustment of the eyes when one passes from a bright to a darkened area.

Defect: A discontinuity that interferes with the usefulness of an object. A fault in a material or test object that is detrimental to its serviceability.

Detergent remover: A penetrant remover that is a solution of a detergent in water. Also see **Hydrophilic emulsifier**.

Developer: A material that is applied to the test object surface after excess penetrant has been removed and that is designed to enhance the penetrant bleedout to form indications. The developer may be a fine powder, a solution that dries to a fine powder or a suspension (in solvent, water, alcohol, etc.) that dries leaving an absorptive film on the test surface.

Developing time: The time between the application of the developer and the examination of the test object for indications. The elapsed time necessary for the applied developer to bring out indications from penetrant entrapments. Also called *development time*.

Discontinuity: An interruption in the normal physical structure or configuration of an object, such as cracks, forging laps, seams, inclusions, porosity, etc. A discontinuity may or may not affect the usefulness of the test object.

Dragout: The loss of penetrant materials from a tank as a result of their adherence to the objects being processed.

Drain time: That portion of the penetrant testing process during which the excess penetrant, emulsifier, detergent remover or developer is allowed to drain from the test object.

Dry developer: A fine dry powder developer that does not use a carrier fluid.

Drying oven: An oven used for drying rinse water from test objects.

Drying time: The time allotted for a rinsed test object to dry.

Dual sensitivity penetrant: A penetrant that contains a combination of visible and fluorescent dyes.

Dwell time: The total time that the penetrant or emulsifier is in contact with the test surface, including the time required for application and the drain time. Also see **Emulsification time**.

Electrostatic spraying: A technique of spraying wherein the material being sprayed is given a high electrical charge, while the test object is grounded.

Emulsification time: The period of time that an emulsifier is permitted to combine with the penetrant before removal. Also called *emulsifier dwell time*.

Emulsifier: A liquid that combines with an oily penetrant to make the penetrant water washable. Also see **Hydrophilic emulsifier** and **Lipophilic emulsifier**.

Evaluation: The process of determining the severity of the condition after the indication has been interpreted. Evaluation leads to determining whether the object is acceptable, salvageable or rejectable.

False indication: An indication caused by improper processing. Distinct from nonrelevant indication.

Flash point: The lowest temperature at which a volatile flammable liquid will give off enough vapor to make a combustible explosive mixture in the air space surrounding the liquid surface.

Flaw: See **Discontinuity**.

Fluorescence: The emission of visible radiation by a substance as the result of, and only during, the absorption of ultraviolet radiation.

Fluorescent penetrant: A testing penetrant that is characterized by its ability to fluoresce when excited by ultraviolet radiation.

Hydrophilic emulsifier: A water based agent that, when applied to an oily penetrant, renders the penetrant water washable. Can be used as a contact emulsifier, but more often the emulsifier is added to the water rinse and accompanied by some form of mechanical agitation or scrubbing to remove excess penetrant. Sometimes called a *hydrophilic remover*.

Indication: That which marks the presence of a discontinuity, as the result of detectable bleedout of penetrant from the discontinuity.

Inspection: The visual examination of a test object after completion of the penetrant processing steps.

Interpretation: The determination of the significance of indications from the standpoint of whether they are relevant or nonrelevant.

Leak testing: A technique of liquid penetrant testing in which the penetrant is applied to one side of the surface, while the other side is tested for indications that would indicate a leak or void.

Lipophilic emulsifier: An oil based agent that, when applied to an oily penetrant, renders the penetrant water washable. Usually applied as a contact emulsifier.

Nonaqueous wet developer: A developer in which the developing powder is applied as a suspension in a quick drying solvent. Also called *solvent developer*.

Nonfluorescent penetrant: See **Visible dye penetrant**.

Nonrelevant indication: An indication that is not or cannot be associated with a discontinuity.

Penetrability: The property of a penetrant that causes it to find its way into very fine openings, such as cracks.

Penetrant: A liquid capable of entering discontinuities open to the surface, and which is adapted to the testing process by being made highly visible in small traces. Fluorescent penetrants fluoresce brightly under ultraviolet radiation, while the visible penetrants are intensely colored to be noticeable under visible light.

Penetration time: See **dwell time**.

Postemulsification: A penetrant removal technique using a separate emulsifier.

Postemulsifiable penetrant: A penetrant that requires the application of a separate emulsifier to render the surface penetrant water washable. Can be removed by applying a solvent remover.

Precleaning: The removal of surface contaminants or smeared metal from the test object so that they cannot interfere with the penetrant testing process.

Quenching of fluorescence: The extinction of fluorescence by causes other than removal of ultraviolet radiation (the exciting radiation).

Rinse: The process of removing liquid penetrant testing materials from the surface of a test object by washing or flooding with another liquid – usually water. Also called *wash*.

Self emulsifiable: See **Water washable penetrant**.

Sensitivity: The ability of the penetrant process to detect surface discontinuities.

Solvent developer: See **Nonaqueous wet developer**.

Solvent removed: A penetrant removal technique wherein the excess penetrant is wiped from the test surface with a solvent remover.

Solvent remover: A volatile liquid used to remove excess surface penetrant from the test object. Sometimes called *penetrant remover*.

Surface tension: That property of liquids which, because of molecular forces, tends to bring the contained volume into a form having the least superficial area.

System: With respect to liquid penetrant testing materials, a combination of liquid penetrant and emulsifier that are furnished by the same manufacturer and are qualified together. For water washable and solvent removable liquid penetrants, a system consists of the liquid penetrant only.

Ultraviolet radiation (for liquid penetrant testing): Light radiation in the near ultraviolet range (UV-A) of wavelengths (320 to 400 nm), just shorter than visible light.

Ultraviolet radiation filter: A filter that transmits near ultraviolet radiation while suppressing visible light and harmful ultraviolet radiation.

Visibility: The characteristic of an indication that enables the observer to see it against the conditions of background, outside light, etc.

Viscosity: The state or degree of being viscous. The resistance of a fluid to the motion of its particles.

Visible dye penetrant: A testing penetrant that is characterized by its intense visible color – usually red. Also called *color contrast or nonfluorescent penetrant*.

Wash: See **Rinse**.

Water soluble developer: A developer in which the developer powder is dissolved in a water carrier to form a solution. Not a suspension.

Water suspendible developer: A developer in which the developer particles are mixed with water to form a suspension.

Water wash: A penetrant removal technique wherein excess penetrant is washed or flushed from the test surface with water.

Water washable penetrant: A type of penetrant that contains its own emulsifier, making it water washable.

Water tolerance: The amount of water that a penetrant, emulsifier can absorb before its effectiveness is impaired.

Wet developer: A developer in which the developer powder is applied as a suspension or solution in a liquid – usually water or alcohol.

Wettability: The ability of a liquid to spread out spontaneously and adhere to solid surfaces.

Bibliography and Figure Sources

BIBLIOGRAPHY

1. *Nondestructive Testing Handbook*, third edition: Vol. 2, *Liquid Penetrant Testing*. Columbus, Ohio: American Society for Nondestructive Testing (1999).

2. *Nondestructive Testing Classroom Training Handbook*, fourth edition: *Liquid Penetrant Testing*. Fort Worth, Texas: Convair Division of General Dynamics Corporation (1981).

FIGURE SOURCES

The following credits indicate the sources of illustrations in this book. All figures reprinted with permission.

Chapter 1

Figure 1.1: Courtesy of Met-L-Chek Company, Santa Monica, California.

Figure 1.2: Reprinted from *Nondestructive Testing Classroom Training Handbook*, fourth edition: *Liquid Penetrant Testing*.

Chapter 2

Figures 2.1 to 2.23: Reprinted from *Nondestructive Testing Classroom Training Handbook*, fourth edition: *Liquid Penetrant Testing*.

Figure 2.24 to 2.26: Reprinted from *Nondestructive Testing Handbook*, third edition: Vol. 2, *Liquid Penetrant Testing*.

Chapter 4

Figures 4.1 to 4.3, 4.8, 4.11: Reprinted from *Nondestructive Testing Classroom Training Handbook*, fourth edition: *Liquid Penetrant Testing*.

Figures 4.4 to 4.6, 4.12, 4.13 : Reprinted from *Nondestructive Testing Handbook*, third edition: Vol. 2, *Liquid Penetrant Testing*.

Figure 4.7: Courtesy of TK Consulting, LLC, Wyoming, Ohio.

Figure 4.9 and 4.10: Courtesy of Magnaflux, Division of Illinois Tool Works, Glenview, Illinois.

Chapter 7

Figure 7.1: Reprinted from *Nondestructive Testing Handbook*, third edition: Vol. 2, *Liquid Penetrant Testing*.

Figure 7.2 to 7.4: Reprinted from *Nondestructive Testing Classroom Training Handbook*, fourth edition: *Liquid Penetrant Testing*.

Index

F

G

H

I

K

L

M

N

O

P

Q

R

S